T0136362

Biotechnological Applications of Microalgae

Biodiesel and Value-Added Products

Biotechnological Applications of Microalgae

Biodiesel and Value-Added Products

Edited by

Faizal Bux

CRC Press
Taylor & Francis Group
Boca Raton London New York

CRC Press is an imprint of the
Taylor & Francis Group, an **informa** business

CRC Press
Taylor & Francis Group
6000 Broken Sound Parkway NW, Suite 300
Boca Raton, FL 33487-2742

First issued in paperback 2019

© 2013 by Taylor & Francis Group, LLC
CRC Press is an imprint of Taylor & Francis Group, an Informa business

No claim to original U.S. Government works

ISBN-13: 978-1-4665-1529-1 (hbk)
ISBN-13: 978-0-367-38002-1 (pbk)

This book contains information obtained from authentic and highly regarded sources. Reasonable efforts have been made to publish reliable data and information, but the author and publisher cannot assume responsibility for the validity of all materials or the consequences of their use. The authors and publishers have attempted to trace the copyright holders of all material reproduced in this publication and apologize to copyright holders if permission to publish in this form has not been obtained. If any copyright material has not been acknowledged please write and let us know so we may rectify in any future reprint.

Except as permitted under U.S. Copyright Law, no part of this book may be reprinted, reproduced, transmitted, or utilized in any form by any electronic, mechanical, or other means, now known or hereafter invented, including photocopying, microfilming, and recording, or in any information storage or retrieval system, without written permission from the publishers.

For permission to photocopy or use material electronically from this work, please access www.copyright.com (http://www.copyright.com/) or contact the Copyright Clearance Center, Inc. (CCC), 222 Rosewood Drive, Danvers, MA 01923, 978-750-8400. CCC is a not-for-profit organization that provides licenses and registration for a variety of users. For organizations that have been granted a photocopy license by the CCC, a separate system of payment has been arranged.

Trademark Notice: Product or corporate names may be trademarks or registered trademarks, and are used only for identification and explanation without intent to infringe.

Library of Congress Cataloging-in-Publication Data

Biotechnological applications of microalgae : biodiesel and value added products / editor,
    Faizal Bux.
      pages cm
      Summary: "The book gives an in-depth analysis of microalgal biology, ecology,
biotechnology and biofuel production capacity as well as a thorough discussion on the value
added products that can be generated from diverse microalgae. It summarizes the state of
the art in microalgal biotechnology research, from microalgal strain selection, microbiology,
cultivation, harvesting, and processing. Contributors from the US, Africa, Asia, South
America, and Europe cover microalgal physiology, biochemistry, ecology, molecular biology,
and more"-- Provided by publisher.
      Includes bibliographical references and index.
      ISBN 978-1-4665-1529-1 (hardback)
      1. Microalgae--Biotechnology. 2. Biomass energy. 3. Microalgae--Industrial use. I. Bux, F.
(Faizel)

   TP248.27.A46B56 2013
   579.8--dc23   2013004139

Visit the Taylor & Francis Web site at
http://www.taylorandfrancis.com

and the CRC Press Web site at
http://www.crcpress.com

Contents

Preface

Over the decades, much of the literature has focused on the biological and ecological aspects of algae found in freshwater, marine, and brackish environments. These organisms are also known to inhabit various other environments on Earth. More recently, there has been a substantial shift toward the concept of sustainable development and the "green economy" with emphasis on exploiting biological systems for the benefit of mankind. This underpins the fundamentals of the field of biotechnology, which has revolutionized various fields including agriculture, food, pharmaceutical and medical sciences, environmental sciences, and industrial feedstock, thus positively impacting most spheres of human endeavor. Algae—but more specifically, microalgae—have been associated with problematic events such as algal blooms caused by eutrophication of aquatic environments, and in some cases the toxins produced have serious health impacts on the aquatic environment, plants, animals, and humans. However, these events are largely due to human activities resulting in the proliferation of nutrients in aquatic environments. The significance of algae cannot be underestimated as they contribute approximately 40% of the oxygen in the atmosphere, are the original source of fossil fuels, and are the primary producers in the oceans. Therefore, there is potential for exploitation of this invaluable biomass source that could lead to definite environmental and economic benefits for man.

Therefore, the present book attempts to encompass the latest developments and recent research thrust and trends focusing specifically on the potential biotechnological application of microalgae. Strain selection, growth characteristics, large-scale culturing, and biomass harvesting have been discussed in detail and critically evaluated. With regard to the benefits of using algal technology, the major areas of progress have been in the fields of biodiesel production using microalgae, screening and production of high-value products from algae, and evaluating carbon dioxide sequestration from flue gas as a climate change mitigation strategy. The latter areas of research are clearly central to a sustainable development approach that is currently attracting global attention. Compared to conventional crops, biodiesel from microalgae has shown superior potential in terms of yields and using wastewater as a substrate. However, critical evaluation of the technology has signaled some limitations with regard to large-scale biomass processing, specifically with regard to harvesting and oil extraction. Globally, much attention is currently focused on overcoming these hurdles to improve the techno-economics and make algal biodiesel a fuel for tomorrow. Findings have shown that a biorefinery approach will certainly aid in improving the economics and large-scale application of the technology. Producing high-value products (health care products, pigments, etc.) specifically using microalgae is a large income-generating industry in many parts of the world, with potential for substantial growth. In addition, algal technology can be used as a cheaper option for the treatment of wastewater streams.

Therefore, this book will serve as an excellent reference for researchers and practitioners, providing a comprehensive outline of the most recent developments and advances in the field of biotechnological applications of microalgae.

I would like to thank all the authors for their contributions to this book.

Professor Faizal Bux
Institute for Water and Wastewater Technology
Durban University of Technology

Acknowledgments

The editor would like to thank the following individuals from the Institute for Water and Wastewater Technology, Durban University of Technology, Durban, South Africa for their assistance in the compilation of this book:

- Taurai Mutanda
- Ismail Rawat
- Abhishek Guldhe

About the Editor

Professor Faizal Bux serves as director at the Institute for Water and Wastewater Technology at the Durban University of Technology. He has 17 years of experience in the fields of water and wastewater treatment, bioremediation and biotransformation of industrial effluents for the production of valuable by-products, and biotechnological applications of microalgae including biodiesel. Professor Bux has a BSc (Honors) in microbiology, an MS in technology (biotechnology), and a doctorate in environmental biotechnology. He has published more than sixty scientific papers in leading Science Citation Index (ISI) journals and contributed to six book chapters, nine technical reports, and more than seventy conference presentations, both nationally and internationally. He also serves as an editor for ISI-listed journals and books. Professor Bux serves on the Management Committee of the International Water Association (IWA) specialist group, Microbial Ecology and Water Engineering. He is a member of many professional bodies and is a Fellow of the Water Institute of Southern Africa. He also serves as a scientific advisor for various NGOs, both in South Africa and internationally, especially with regard to water quality issues.

Contributors

Faizal Bux
Institute for Water and Wastewater
  Technology
Durban University of Technology
Durban, South Africa

Tapan Chakrabarti
CSIR-National Environmental
  Engineering Research Institute
  (NEERI)
Nagpur, India

Vikas S. Chauhan
CSIR-Central Food Technological
  Research Institute (CFTRI)
Mysore, India

Keshav Das
Biorefining and Carbon Cycling
  Program
College of Engineering
The University of Georgia
Athens, Georgia

Sivanesan S. Devi
CSIR-National Environmental
  Engineering Research Institute
  (NEERI)
Nagpur, India

Ravi V. Durvasula
Department of Internal Medicine
Center for Global Health
University of New Mexico School of
  Medicine
and
The Raymond G. Murphy VA Medical
  Center
Albuquerque, New Mexico

Sanniyasi Elumalai
Presidency College
Chennai, India

Melinda J. Griffiths
Centre for Bioprocess Engineering
  Research
University of Cape Town, South Africa

Susan T.L. Harrison
Department of Chemical Engineering
Centre for Bioprocess Engineering
  Research
University of Cape Town, South Africa

Subburamu Karthikeyan
Department of Agricultural
  Microbiology
Tamil Nadu Agricultural College
Directorate of Natural Resource
  Management
Coimbatore, India

Kannan Krishnamurthi
CSIR-National Environmental
  Engineering Research Institute
  (NEERI)
Nagpur, India

Ramanathan Ranjith Kumar
Institute for Water and Wastewater
  Technology
Durban University of Technology
Durban, South Africa

Rajesh Lalloo
CSIR Biosciences
Pretoria, South Africa

Yun Liu
College of Life Science and Technology
Beijing University of Chemical
    Technology
Beijing, China

Dheepak Maharajh
CSIR Biosciences
Pretoria, South Africa

Sandeep N. Mudliar
CSIR-National Environmental
    Engineering Research Institute
    (NEERI)
Nagpur, India

Taurai Mutanda
Institute for Water and Wastewater
    Technology
Durban University of Technology
Durban, South Africa

Terisha Naidoo
CSIR Biosciences
Pretoria, South Africa

Renganathan Rajkumar
Department of Chemical and Process
    Engineering
Faculty of Engineering and Built
    Environment
University Kebangsaan Malaysia
Bangi, Malaysia

Desikan Ramesh
Deparment of Farm Machinery
Agricultural Engineering College
    and Research Institute
Tamil Nadu Agricultural University
Coimbatore, India

Durvasula V. Subba Rao
Center for Global Health, Department
    of Internal Medicine
University of New Mexico School of
    Medicine

and
The Raymond G. Murphy VA Medical
    Center
Albuquerque, New Mexico

Vadrevu S. Rao
Department of Mathematics,
Jawaharlal Nehru Technological
    University Hyderabad
Hyderabad, India

Ismail Rawat
Institute for Water and Wastewater
    Technology
Durban University of Technology
Durban, South Africa

Christine Richardson
Department of Chemical Engineering
Centre for Bioprocess Engineering
    Research
University of Cape Town, South Africa

Ravi Sarada
CSIR-Central Food Technological
    Research Institute (CFTRI)
Mysore, India

Yogesh C. Sharma
Department of Applied Chemistry
Indian Institute of Technology (BHU)
Varanasi, India

Ajam Y. Shekh
CSIR-National Environmental
    Engineering Research Institute
    (NEERI)
Nagpur, India

Rheka Shukla
Biorefining and Carbon Cycling
    Program
College of Engineering
The University of Georgia
Athens, Georgia

Bhaskar Singh
Department of Applied Chemistry
Indian Institute of Technology (BHU)
Varanasi, India

Manjinder Singh
Biorefining and Carbon Cycling
    Program
College of Engineering
The University of Georgia
Athens, Georgia

Zahira Yaakob
Department of Chemical and Process
    Engineering
Faculty of Engineering and Built
    Environment
University of Kebangsaan Malaysia
Bangi, Malaysia

Raju R. Yadav
CSIR-National Environmental
    Engineering Research Institute
    (NEERI)
Nagpur, India

Nodumo Zulu
CSIR Biosciences
Pretoria, South Africa

1 Introduction

Taurai Mutanda
Institute for Water and Wastewater Technology
Durban University of Technology
Durban, South Africa

CONTENTS

1.1 GENERAL OVERVIEW

Microalgae are single-celled, ubiquitous, prokaryotic and eukaryotic primary photosynthetic microorganisms that are taxonomically and phylogenetically diverse. The advanced plant life of today is thought to have evolved from these simple microscopic plant-like entities. In general, the algae are a heterogeneous group of polyphyletic photosynthetic organisms with an estimated 350,000 known species (Brodie and Zuccarella, 2007). There are predominantly two prokaryotic divisions (Cyanophyta and Prochlorophyta) and nine eukaryotic divisions (Glaucophyta, Rhodophyta, Heterokontophyta, Haptophyta, Cryptophyta, Dinophyta, Euglenophyta, Chlorarachniophyta, and Chlorophyta). The biology of microalgae is interesting, and their enigma is due to their wide diversity as well as their plethora of habitats. The biology of microalgae is discussed extensively in Chapter 2 of this book.

Interest in microalgal cultivation is currently blossoming globally for a number of reasons. Microalgae are not extremely fastidious microorganisms but are found in diverse aquatic habitats. Microalgae can be found almost anywhere on Earth, in freshwater, marine, and hyper-saline environments (Williams and Laurens, 2010). The nutritional requirements of a wide array of microalgal strains are known, and the technology for microalgal cultivation is developing at a fast pace. The advent of genetic engineering protocols has brought new vistas to algal molecular systematics. Recently, the general study of microalgae using genomics and molecular biology tools has attained phenomenal dimensions. The sheer number of microalgal strains from extreme environments that are yet to be discovered and identified is enormous (Brodie and Lewis, 2007). However, microalgal culture collection banks have been established as repository centers for these microorganisms (e.g., UTEX at The University of Texas at Austin).

The importance of microalgae in day-to-day life cannot be overemphasized. As the main primary producers, microalgal biomass is used for food and feed supplements (Lewis et al., 2000). Microalgae are important sources of commercial products such as polyunsaturated fatty acid (PUFA) oils (e.g., γ-linolenic acid (GLA), arachidonic acid (AA), eicosapentaenoic acid (EPA), and docosahexaenoic acid (DHA)) (Spolaore et al., 2006). In addition, microalgae such as *Dunaliella* and *Haematococcus* are important sources of carotenoids such as β-carotene and astaxanthin, respectively (Spolaore et al., 2006). Furthermore, the cyanobacterium *Anthrospira* and the rhodophyte *Porphyridium* are the main commercial producers of phycobiliproteins (i.e., phycoerythrin and phycocyanin), which are used as natural dyes and for pharmaceutical applications (Spolaore et al., 2006). Potential biotechnological applications and value-added products generated from microalgae are discussed in Chapter 10 of this volume.

Chlorella, Arthrospira, and *Nostoc* are cultivated worldwide for human and animal nutrition, owing to their chemical composition (Spolaore et al., 2006). Microalgae have been hailed as the panacea for the dwindling petroleum-based fuels, and the preponderance of shorter-chain fatty acids has significance for their potential as diesel fuels (Chisti, 2007; Williams and Laurens, 2010). The efficacy of using microalgal biomass and lipids as alternative biofuels is currently a topical issue. Biofuels such as biodiesel, biomethane, biohydrogen, biobutanol, etc., can be generated from microalgae (Chisti, 2007). Current research is targeting other novel potential biotechnological applications in aquaculture, cosmetics, pharmaceuticals, and animal and human nutrition. It is envisaged that future research should focus on microalgal strain improvement through genetic engineering, in order to diversify and economically improve product competitiveness (Spolaore et al., 2006). Microalgal genetic manipulation is still in its infancy and is a pertinent area of investigation in order to improve the quality and quantity of products generated from microalgae. However, the development of nondestructive product recovery techniques from continuous cultivation systems will greatly improve product yield.

Successful microalgal cultivation and generation of these products calls for meticulous and rigorous microalgal strain selection. Two important steps in obtaining a robust and suitable microalgal candidate are (1) bioprospecting of target microalgal strain samples from diverse habitats, and (2) strain selection, isolation, and purification using conventional and advanced microbiological methods (Grobbelaar, 2009; Mutanda et al., 2011). Suitable microalgal strains can be obtained commercially from registered authentic culture collection centers. The microalgal strain of choice is maintained under laboratory conditions, either as a freeze-dried sample or as a slant on solid media at 4°C with routine subculturing. The ever-growing field of phycology has introduced new, exciting, and efficient techniques for maintaining microalgal cultures at ultra-low temperatures (i.e., cryopreservation). Microalgal strain selection for biodiesel production is discussed in detail in Chapter 3 of this volume.

The enumeration of microalgae poses a real challenge due to the requirement of sophisticated equipment such as flow cytometers. The use of optical microscopes for cell counting is relatively cheaper, although not very accurate as compared to faster automated cell counting techniques (Guillard and Sieracki, 2005; Marie et al., 2005). Microalgal cells are counted in order to estimate the size of the cultured population and to estimate the rate of culture growth (i.e., determination of the rate of population increase) (Guillard and Sieracki, 2005). Microalgal enumeration methods are described in detail in Chapter 4 of this volume.

The important factors affecting microalgal growth are light intensity, temperature, nutrients, CO_2 availability, pH, and salinity (Bhola et al., 2011; Rosenberg et al., 2011). Other factors such as conductivity, oxidation/reduction potential (ORP), total dissolved solids (TDS), and biological factors such as protozoa are also important. These factors must be closely monitored to prevent failure of the cultivation system, especially when growing microalgae on a large commercial scale.

There are essentially two commonly used methods for microalgal cultivation, namely open raceway ponds and photobioreactors. The design, and the pros and cons, of these cultivation systems are discussed in detail in Chapter 5. The open raceway system is amenable to large-scale microalgal cultivation because it is simple and cost effective to operate. Despite these attractive features, microalgal biomass harvesting still remains a huge challenge. Harvesting microalgal biomass is technically difficult because the biomass exists as a dilute aqueous suspension. Furthermore, microalgal cells are very difficult to remove due to their miniscule size (<20 µm), similar in density to water (Lavoie and De la Noue, 1986), and strong negative surface charge, particularly during exponential growth (Moraine et al., 1979; Park et al., 2011). It is a relatively daunting task to surmount these drawbacks.

Several methods are available for dewatering and recovering microalgal biomass, such as centrifugation, flocculation, gravity settling, microfiltration, and dissolved air floatation (DAF) *inter alia* (Lavoie and de la Noue, 1986; Molina Grima et al., 2003). The technology for microalgal biomass harvesting is still in its infancy, and trials on suitable combinations of these methods are currently underway (Williams and Laurens, 2010). The use of the centrifugation technique on a large scale is not cost effective due the colossal amounts of power consumption (Mutanda et al., 2011). The techniques available for microalgal harvesting and dewatering are discussed at length in Chapter 6.

There are several techniques that are used for extracting lipids from microalgal biomass (Lewis et al., 2000). Most of these methods are destructive; however, it is desirable to develop nondestructive methods for continuous extraction of lipids from live microalgal cells. The solvent extraction system using a mixture of solvents such as hexane and methanol are commonly used. Other methods are sonication and microwave-assisted extraction. The Bligh and Dyer method (1959) has been commonly used in many applications, whereby lipids are extracted from biological material using a combination of chloroform and methanol (Lewis et al., 2000). Extracting lipids from microalgal biomass is a real challenge because it is intracellular and therefore requires a cell disruption step. Currently, research is ongoing to develop cost-effective and efficient lipid extraction strategies (Molina-Grima et al., 2003; Williams and Laurens, 2010). Subsequent to lipid extraction, it is desirable to accurately identify the lipid and characterize the lipids using highly analytical techniques. This is done to establish whether the lipids extracted are suitable for application to biodiesel production. Techniques that are widely used for the analysis of lipids are gas chromatography with mass spectrometry (GC-MS), liquid chromatography (LC), matrix-assisted laser desorption/ionization–time of flight (MALDI-TOF), thin-layer chromatography (TLC), etc. Chapter 7 explores in detail the lipid extraction and identification techniques that are commonly used.

Microalgal lipids are converted into biodiesel through transesterification steps. Transesterification of microalgal lipids into biodiesel is accomplished either chemically or biologically using lipolytic enzymes. These methods are outlined in Chapter 8. To establish the feasibility of biodiesel production from microalgae, it is prudent to perform a life cycle analysis (LCA). The procedures involved in LCA are discussed in Chapter 9. Apart from generating biofuels and other value-added products, microalgae cultivation is also profoundly involved in climate change abatement through CO_2 sequestration. This important application of microalgae is discussed in Chapter 11. Microalgae can use wastewater rich in nitrates and phosphates as substrates for growth while simultaneously removing these macronutrients and thereby arresting eutrophication. Therefore, microalgae are involved in the phycoremediation of domestic and industrial wastewaters, and this is achieved in high-rate algal ponds (Chapter 12). Finally, Chapter 13 discusses general microalgal biotechnology in terms of its potential as today's "green gold rush." The chapter gives an overview of advanced techniques such as genetic engineering of microalgae so as to increase lipid yield.

ACKNOWLEDGMENTS

The author hereby acknowledges the National Research Foundation (South Africa) for financial assistance.

REFERENCES

Bhola, V., Ramesh, D., Kumari-Santosh, S., Karthikeyan, S., Elumalai, S., and Bux, F. (2011). Effects of parameters affecting biomass yield and thermal behavior of *Chlorella vulgaris*. *Journal of Bioscience and Bioengineering*, 111: 377–382.

Bligh, E.G., and Dyer, W.J. (1959). A rapid method for total lipid extraction and purification. *Canadian Journal of Biochemistry and Physiology*, 37: 911–917.

Brodie, J., and Lewis, J. (2007). *Unravelling the Algae, the Past, Present, and Future of Algal Systematics*. CRC Press, London.

Brodie, J., and Zuccarello, G.C. (2007). Systematics of the species-rich algae: Red algal classification, phylogeny and speciation. In *The Taxonomy and Systematics of Large and Species-Rich Taxa: Building and Using the Tree of Life* (Eds. T.R. Hodkinson and J. Parnell), Systematics Association Series, CRC Press, London, pp. 317–330.

Chisti, Y. (2007). Biodiesel from microalgae. *Biotechnology Advances*, 25: 294–306.

Grobbelaar, J.U. (2009). From laboratory to commercial production: A case study of a *Spirulina* (*Arthrospira*) facility in Musina, South Africa. *Journal of Applied Phycology*, 21: 523–527.

Guillard, R.R.L., and Sieracki, M.S. (2005). Counting cells in cultures with the light microscope. In R.A. Andersen (Ed.). *Algal Culturing Techniques* (pp. 239–252). Elsevier Academic, Burlington, MA.

Lavoie, A., and De la Noue, J., (1987). Harvesting of *Scenedesmus obliquus* in wastewaters: Auto- or bioflocculation. *Biotechnology and Bioengineering*, 30: 852–859.

Lewis, A., Nichols, P.D., and McMeekin, T.A. (2000). Evaluation of extraction methods for recovery of fatty acids from lipid-producing microheterotrophs. *Journal of Microbiological Methods*, 43: 107–116.

Marie, D, Simon, N., and Vaulot, D. (2005). Phytoplankton cell counting by flow cytometry. In R.A. Andersen (Ed.). *Algal Culturing Techniques* (pp. 253–268). Elsevier Academic, Burlington, MA.

Molina Grima, E., Belarbi, E.H., Acien Fernandez, F.G., Robles Medina, A., and Chisti, Y. (2003). Recovery of microalgal biomass and metabolites: Process options and economics. *Biotechnology Advances,* 20: 491–515.

Moraine, R., Shelef, G., Meydan, A., and Levi, A., (1979). Algal single cell protein from wastewater treatment and renovation process. *Biotechnology and Bioengineering*, 21: 1191–1207.

Mutanda, T., Ramesh, D., Karthikeyan, S., Kumari, S., Anandraj, A., and Bux, F. (2011). Bioprospecting for hyper-lipid producing microalgal strains for sustainable biofuel production. *Bioresource Technology,* 102: 57–70.

Park, J.B.K., Craggs, R.J., and Shilton, A.N. (2011). Wastewater treatment high rate algal ponds for biofuel production. *Bioresource Technology,* 102: 35–42.

Rosenberg, J.N., Mathias, A., Korth, K., Betenbaugh, M.J., and Oyler, G.A. (2011). Microalgal biomass production and carbon dioxide sequestration from an integrated ethanol biorefinery in Iowa: A technical appraisal and economic feasibility evaluation. *Biomass and Bioenergy,* 35: 3865–3876.

Spolaore, P., Joannis-Cassan, C., Duran, E., and Isambert, A. (2006). Commercial applications of microalgae. *Journal of Bioscience and Bioengineering,* 101, 87–96.

Williams, P.J.B., and Laurens, L.M.L. (2010). Microalgae as biodiesel and biomass feedstocks: Review and analysis of the biochemistry, energetics and economics. *Energy and Environmental Science,* 3: 554–590.

2 The Biology of Microalgae

Ranganathan Rajkumar and Zahira Yaakob
Department of Chemical and Process Engineering
University of Kebangsaan, Bangi, Malaysia

CONTENTS

2.1   TAXONOMY

Algae are a diverse group of organisms that can perform photosynthesis efficiently. On the basis of morphology and size, algae can be subdivided and are classified into two main categories: macroalgae and microalgae. Macroalgae consist of multiple cells that organize into structures resembling the roots, stems, and leaves of higher plants (e.g., kelp). Microalgae are an extremely diverse group of primary producers present in almost all ecosystems on Earth, ranging from marine, freshwater, desert sands, and hot springs, to snow and ice (Guschina and Harwood, 2006). They are colonial or single-celled organisms that have garnered increasing amounts of attention and interest for industrial purposes. They are categorized into divisions based on various characteristics such as morphological features, pigmentation, the chemical nature of photosynthetic storage products, and the organization of photosynthetic membranes. The four most important algal groups in terms of abundance are green algae (Chlorophyceae), diatoms (Bacillariophyceae), blue-green algae (Cyanophyceae), and golden algae (Chrysophyceae) (Khan et al., 2009). According to estimations reported by Cardozo et al. (2007), they include between 200,000 and 800,000 species, of which about only 35,000 species have been described.

2.1.1   GENERAL CHARACTERISTICS

The main characteristics of microalgae are primarily simple morphological features that can easily be observed under a light microscope. Cyanophyceae consist of prokaryotic cells commonly called the blue-green algae. They are similar to Gram-negative bacteria, based on the nature of the cell wall, cell structure, and capacity to fix atmospheric nitrogen; hence they are called cyanobacteria. However, they possess the photosynthetic system chlorophyll-*a*, accessory pigments, and thallus organization that resembles other algae. Cyanophyceae members can be broadly divided into coccoid forms and filamentous forms. The coccoid has various forms, from single cell to aggregates of unicellular cells; regular or irregular colonies; and pseudofilamentous and pseudoparenchymatous conditions. The filamentous forms exist as simple uniseriate filaments to heterotrichous filaments, which may be differentiated into heterocysts and akinetes. These are ubiquitous in nature, occurring in several habitats with extreme conditions (i.e., temperature, light, pH, and nutritional resources). They are found abundantly in a variety of natural and artificial aquatic ecosystems. Cyanophyceae members can be easily identified within a mixture of other algae by their distinct blue-green color.

The Chlorophyceae constitute a major group of algae occurring in various habitats. The cells are usually green in color due to the presence of pigments such as chlorophyll-*a* and -*b*. The cells contain chloroplasts of various shapes that are located differently in each group of organisms. In addition, the chloroplasts also contain pyrenoids. The nucleus of this group may present either singly or in multiples but in major organisms occurs singly although some genera are multinucleate. Flagellated cells are common either in the vegetative or reproductive phase. There is at least one group without any flagellated cells.

Euglenophyceae members are unicellular, motile, and usually contain one prominent flagellum and in some cases two flagella. The anterior position of a cell has a visible gullet, and many dissimilar chloroplasts are found in autotrophic forms and are absent in other forms. Euglenoid cells are enclosed by a proteinacious pellicle and help the organisms achieve pleomorphism. These are widely distributed in all types of water bodies, particularly in organic-rich aquatic ecosystems.

Bacillariophyceae members are popularly known as diatoms. They are basically unicellular, and also occur as pseudofilaments or aggregated in colonies. The cell wall of a diatom is impregnated with silica and they have been well preserved as microfossils. The diatom cell is also called a frustule, and the classification of diatoms is based on the pattern of ornamentation on their wall. The cells have either radial or bilateral symmetry. The frustules consist of two halves (epitheca and hypotheca) and connecting girdle bands. The surface of the valve has typical markings. Punctae are regularly or irregularly arranged to form striae. Areolae are pores or chambers within the valve wall. Costae are elongated thickenings of the valve wall due to heavy deposition of silica. The valves of some diatoms have an opening or fissure along the apical axis called the raphe. The presence of the raphe or its absence on the walls of the diatoms has been one of the features in the identification of diatoms and distinguishing different genera. The radially symmetric forms are grouped as Centracles and the bilaterally symmetric ones are Pennales.

2.2 MORPHOLOGICAL IDENTIFICATION

Understanding biodiversity is critical in ecological research because it unravels the role of each single species in the ecosystem in mediating the environment for the entire biological community. The microalgal biodiversity of a region has economic value and, hence, any loss in biodiversity is of serious economic concern. This emphasizes the need and importance of biodiversity conservation. A survey of the literature on microalgae diversity during the past four decades has revealed that ecosystems harbor a large number of algae belonging to various groups. Despite the availability of elaborate monographs on specific groups of algae, any significant amounts of literature on the taxonomy of several algal species and genera remain scarce and scattered.

Proper identification of the algal taxa has always been considered "not an easy task." Biochemical investigations of, and research into, environmentally important organisms have diversified phycological research and allowed it to enter a whole new phase altogether. Discovery of the potential of microalgae for industrial production of certain chemicals of pharmaceutical value led to the involvement of multifarious types of scientists in understanding the algae. Researchers on the threshold of algal taxonomic study and nonbotanists who are intrigued to know about the algae for pursuing research of their own interest are overwhelmed by the huge literature available. The present work therefore is an attempt to bring together a basic way of identifying all microalgae belonging to many of the genera of various groups of algae that occur abundantly and commonly in all ecosystems. Photographs were prepared for various groups of algae as described here. These photographs contain taxonomic information on the individual groups of organisms, especially the identification of genera and species. Different species of microalgae were recorded from water samples. All species of microalgae belonged to two major groups of algae, namely Bacillariophyceae and Dinophyceae. The Bacillariophyceae members were represented by *Mastogloia paradoxa* Grun., *Rhabdonema adriatium* Ktz., *Synedra gruvei* Grunow, *Chaetoceros orientalis* Schiller, *Nitzschia draveillensis* Coste & Ricard, *Pleurosigma formosum* Wm. Smith, *Coscinodiscus janischii* var. *arafurensis* Grun., *Cocconeis scutellum* Ehrenb., *Podocystis spathulata* (Shadbolt) Van Heurck, *Actinocylus octonarius* Ehrenb., *Biddulphia biddulphiana* (Smith) Boyer, *Thalassionema nitzshioides* Grun., *Rhizosolenia setigera* Brightwell, and *Thalassiothrix longissima* Cleve & Grun. The Dinoflagellates are represented by *Ceratium hirundinella* (Muller) Dujardin, *Ceratium longipeps* (Bailey) Grun., *Ceratium trichoceros* (Ehrenberg) Kofoid, and *Gymnodinium sanguineum* Hirasaka (Figures 2.1, 2.2, 2.3, and 2.4; see color insert) (Rajkumar, 2010).

2.3 MOLECULAR IDENTIFICATION

Recent research in microalgal ecology, physiology, systematics, and genomics has revealed a vast, unexpected diversity. The estimation of microalgal biodiversity has been hindered by cultivating microalgae for commercial products. Molecular identification serves as a prominent tool to distinguish inter- and intra-specific morphologically similar species (Olmos et al., 2000) and mixed populations (Olsen et al., 1986). The developments of modern biotechnological tools, such as polymerase chain reaction (PCR)- and rDNA-based technologies facilitate in detecting small numbers of microalgae in complex natural populations and are widely applied to ascertain the systematic position of species. Sequence analysis has been used to clarify the taxonomic affinities of a wide range of taxa (McInnery et al., 1995;

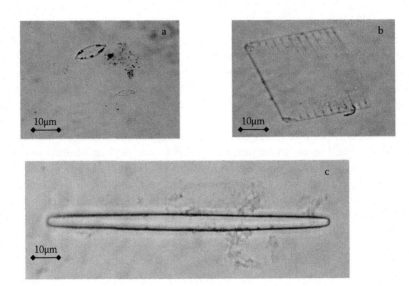

FIGURE 2.1 (See color insert.) Morphological diversity of microalgae: (a) *Mastogloia paradoxa* Grun., (b) *Rhabdonema adriatium* Ktz., (c) *Synedra gruvei* Grun.

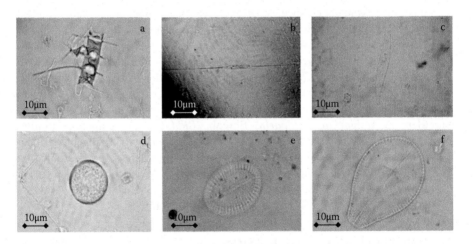

FIGURE 2.2 (See color insert.) Morphological diversity of microalgae: (a) *Chaetoceros orientalis* Schiller, (b) *Nitzschia draveillensis* Coste & Ricard, (c) *Pleurosigma formosum* Wm. Smith, (d) *Coscinodiscus janischii* var. *arafurensis* Grun., (e) *Cocconeis scutellum* Ehrenb., (f) *Podocystis spathulata* (Shadbolt) Van Heurck.

Baker et al., 1999) and as a powerful tool for assessing the genetic diversity of environmental samples (Van Waasberngen et al., 2000; Baker et al., 2001). Apart from detecting genetic diversity, molecular tools may also help in detecting the spatial repartition of an organism, both in marine and freshwater microalgae. PCR-based methods are more commonly utilized due to their rapid results. They have been successfully used to detect the genetic make-up of various natural samples.

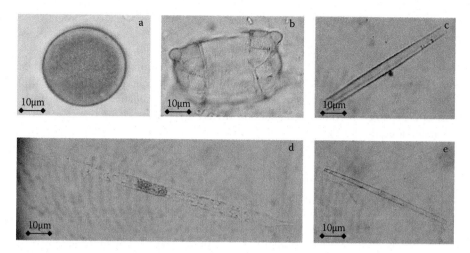

FIGURE 2.3 (See color insert.) Morphological diversity of microalgae: (a) *Actinocylus octonarius* Ehrenb., (b) *Biddulphia biddulphiana* (Smith) Boyer, (c) *Thalassionema nitzshioides* Grun., (d) *Rhizosolenia setigera* Btw., (e) *Thalassiothrix longissima* Cleve & Grun.

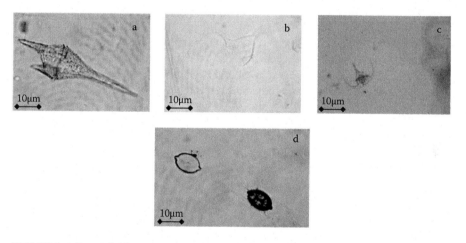

FIGURE 2.4 (See color insert.) Morphological diversity of microalgae: (a) *Ceratium hirundinella* (Muller) Dujardin, (b) *Ceratium longipeps* (Bailey) Grun., (c) *Ceratium trichoceros* (Ehrenberg) Kofoid, (d) *Gymnodinium sanguineum* Hirasaka.

However, a problem with several commonly used primers is that they are constructed theoretically and from an incomplete database of 18S rRNA sequences from cultured organisms. Therefore, experiential testing is pivotal to confirm PCR primer specificity prior to their use in environmental samples. Thus far, a large number of primer sequences have been produced for amplification and sequencing of ssu-rRNA genes. Some of them have been designed as taxa specific, whereas others, designed to amplify all prokaryotic ssu-rRNA genes, are referred to as universal primers. Because the database of 18S rRNA gene sequences has grown, new taxonomic groups have been revealed. Formal analyses of species borders have been made possible due to modern advances in techniques for sequence-based species delimitation (Wiens, 2007; Zhang et al., 2008). A variety of methods for detecting species are based on analytical character variation limits from DNA sequence data. Hence, these methods are rooted in phylogenetic species insight, aggregating a population that lacks separate variations into a single species, and distinguishing other species by distinct nucleotide differences (Wiens and Penkrot, 2002; Monaghan et al., 2005). Among these methods, statistical parsimony (Templeton et al., 1992) segregates a group of sequences if genotypes are connected by long branches that are affected by homoplasy. In recent times, the maximum likelihood approaches aim to connect between statistics and sequence data by analyzing the dynamics of lineage branching in phylogenetic trees for determining the species boundaries. This technique attempts to determine the point of transition from species-level (speciation) to population-level (coalescent) evolutionary processes (Pons et al., 2006; Fontaneto et al., 2007). The similar morphological or polymorphic traits of many algal groups that can be achieved by sequence-based identification are, in fact, valuable discoveries in this field (Harvey and Goff, 2006; Lilly et al., 2007; Vanormelingen et al., 2007). The taxonomy of the most common microalgal species is sorted using available distributional information that is reliable. The distribution records, along with physiological information, will allow for designing ecological models incorporating the special effects of climatic parameters, which in turn would be very useful to predict transfer in distribution due to climatic changes. At present, such models are nonexistent for microalgae. As a general conclusion, the biogeography of algae is a poorly explored area, but holds great potential for exciting research and is definitely worthy of much greater interest than received thus far.

2.4 HABITATS

Microalgal biodiversity or the variation in life forms within a given ecological region is often considered the measure of a specified ecosystem. Biodiversity indicates not only the species abundance of the area, but also the sum of the genera, species, and ecosystems of a region. The water bodies in these diverse habitats harbor a wide variety of microalgae that appear, disappear, and reappear during the changing seasons.

2.4.1 Freshwater

Algae present in various freshwater habitats such as ponds, puddles, lakes, agricultural lands, oxidation ponds, streams, canals, springs, water storage tanks, reservoirs, and rivers are described and enumerated. Sub-aerial algae that transpire on moist tree barks, the walls of buildings, and dripping rocks are also considered freshwater algae by various algologists. The information on freshwater algae is vast, yet remains scattered. Blue-green algae, green algae, diatoms, and euglenoid flagellates are the main components of freshwater habitats. There are also other types, such as planktons (free floating), benthons (attached to sediments), epiphytic algae (attached to larger algae and hydrophytes), epilithic algae (on stones and rocks of reservoirs and lakes), epipelic algae (attached to sand and mud), endophytes (living inside the tissue of other plants), epizoic algae (on shell of snails), and endozoic forms (inside sponges). Considerable research has been carried out on various water bodies with reference to hydrological conditions and the periodicity, abundance, and seasonal variation of algae. Members of the green algae dominate in summer and become less dominant in winter. High temperatures and organic matter support the occurrence of euglenoids, while high phosphates and low organic matter facilitate the abundance of diatoms. An increase in pH, organic matter, and nutrients sometimes leads to the formation of blooms. *Microcystis aeruginosa* forms blooms in several highly alkaline and nutrient-rich ponds. Diurnal variations of microalgae have also been noticed. Lakes and reservoirs have been found to show highest algae occurrence during summer and winter. The shallow waters near the shores contain epiphytic filamentous green algae, and the deeper waters support only deep euplanktonic organisms such as desmids and diatoms. Blue-green algae occur in low dissolved oxygen, abundant organic matter, and high temperature, and become possible indicators of pollution, while green algae occur with exactly the opposite conditions. Diatoms show seasonal periodicity, always reaching a maximum during summer, and correlate with the silicates. Desmids are very sensitive to pollution. Hydrobiological studies on polluted waters have led to spotting microalgal indicators by calculating the species diversity index. Studies on the relationship of microalgae and their nutrient requirements in some lakes have in certain cases led to developing approaches for conservation and prevention from pollution.

2.4.2 Marine

Marine ecosystems that form the marine environment are the largest habitats on Earth for a diversified group of organisms. The biotic community of marine environments is dominated by microalgae. They are among the largest primary producers of biomass in the marine environment and are common inhabitants of the tidal and intertidal areas of the marine ecosystem. These algae exhibit a characteristic geographical distribution pattern under the influence of several environmental factors (Vijayaraghavan and Kaur, 1997). Nevertheless, the coastal ecosystem in the marine environment is very complex where all organisms exist in mutual dependency. Although considerable attention has been paid to distribution, abundance, growth, culture, biochemical constituents, by-products, and bioactivity of marine algae in different parts of the world (Faulkner, 1984), information is available on the microbes, planktonic and faunal associates of marine algae, and the impact of various environmental factors on their distribution. The algal biotope with its morphological diversity is considered important in providing food, living space, and refuge, and offers a variety of potential habitats for the faunal species, including planktonic forms. A detailed investigation is therefore necessary to understand the actual nature of the association between algae and other forms to appreciate the potential importance of this interaction in the marine ecosystem. Steele (1988) described the "heuristic projection," which illustrates the scales of importance in monitoring pelagic components of the ecological unit. The large marine ecosystem (LME) approach defines a spatial domain based on ecological principles and, thereby, provides a basis for focused temporal and spatial scientific research and monitoring efforts in support of management aimed at the long-term productivity and sustainability of marine habitats and resources. The plankton of LMEs can be studied by deploying Continuous Plankton Recorder (CPR) systems (Glover, 1967) through commercial vessels. Advanced plankton recorders can be installed with sensors for intense recording of temperature, salinity, chlorophyll, nitrate/nitrite, petroleum hydrocarbons, light, bioluminescence, and primary productivity (UNESCO, 1992; Williams, 1993), which will help monitor the changes in phytoplankton composition, dominance, and long-standing changes in the physical and nutrient characteristics of the LME. In addition, longer-term changes in relation to the biofeedback of the plankton community toward adverse climate may also be clearly understood (Hayes et al., 1993; Jossi and Goulet, 1993; Williams, 1993).

The phytoplankton community includes 5,000 marine species of unicellular algae and has a broad diversity of cell size (mostly in the range of 1 to 100 μm), morphology, physiology, and biochemical composition (Margalef, 1978). All phytoplankton species are capable of photosynthesis, and many have the capacity for rapid cell division and population growth—up to four doublings per day. The population dynamics of the phytoplankton can be interpreted as responses to changes in the individual processes that regulate the biomass (total quantity, in measures such as carbon, nitrogen, or chlorophyll concentration), species composition, and spatial distribution of the phytoplankton population. Phytoplankton have a wide distribution in all habitats of the marine environment and play a major role in the food chain of an aquatic ecosystem. Some of the phytoplankton species also act as bio-indicators, reflecting changes in the environment. Different hydrobiological parameters, such as pH, temperature, salinity, alkalinity, nutrient concentration, solar radiation, etc., determine species composition, diversity, succession, and abundance of phytoplankton (Perumal et al., 1999; Redekar and Wagh, 2000a, b). Remarkable changes in the irradiance toward phytoplankton could occur due to changing seasonal, diurnal cycles and weather conditions. Diatoms are the significant and often dominant constituent of benthic microalgal communities in estuarine and shallow coastal regions (Sullivan, 1999).

The taxonomy of the most common species with reliable distributional information and records will allow for the design of ecological role models incorporating the effects of climatic parameters, which would be very useful in predicting shifts in distribution due to climatic changes.

REFERENCES

Baker, G.C., Beebee, T.J.C., and Ragan, M.A. (1999). *Prototheca richardsi*, a pathogen of anuran larvae, is related to a clade of protistan parasites near the animal-fungal divergence. *Microbiology*, 145: 1777–1784.

Baker, G.C., Gaffar, S., Cowan, D.A., and Suharto, A.R. (2001). Bacterial community analysis of Indonesian hot springs. *FEMS Microbiology Letters*, 200: 103–109.

Cardozo, K.H., Guaratini, T., Barros, M.P., et al. (2007). Metabolites from algae with economical impact. *Comparative Biochemistry and Physiology - Part C: Toxicology & Pharmacology*, 146: 60–78.

Faulker, D.J. (1984). Marine natural products: Metabolites of marine algae and herbivorous marine mollusks. *Natural Products Report,* 1: 251–280.

Fontaneto, D., Herniou, E.A., Boschetti, C., Caprioll, M., Melone, G., Ricci, C., and Barraclough, T.G. (2007). Independently evolving species in asexual bdelloid rotifers. *PLoS Biology,* 5: 914–921.

Glover, R.S. (1967). The Continuous Plankton Recorder Survey of the North Atlantic. *Symposia of the Zoological Society of London,* 19: 189–210.

Guschina, I.A., and Harwood, J.L. (2006). Lipids and lipid metabolism in eukaryotic alga. *Progress in Lipid Research,* 45: 160–186.

Harvey, J.B.J., and Goff, L.J.A. (2006). A reassessment of species boundaries in *Cystoseira* and *Halidrys* (Phaeophyceae, Fucales) along the North American west coast. *Journal of Phycology,* 42: 707–720.

Hayes, G.C., Carr, M.R., and Taylor, A.H. (1993). The relationship between Gulf Stream position and copepod abundance derived from the Continuous Plankton Recorder survey: Separating the biological signal from sampling noise. *Journal of Plankton Research,* 15: 1359–1373.

Jossi, J.W., and Goulet, J.R. (1993). Zooplankton trends: U.S. north-east shelf ecosystem and adjacent regions differ from north-east Atlantic and North Sea, ICES. *Journal of Marine Science,* 50: 303–313.

Khan, S.A., Rashmi, M.Z., Hussain, Prasad, S., and Banerje, U.C. (2009). Prospects of bio-diesel production from microalgae in India. *Renewable and Sustainable Energy Reviews,* 13(9): 2361–2372.

Lilly, E.L., Halanych, K.M., and Anderson, D.M. (2007). Species boundaries and global biogeography of the *Alexandrium tamarense* complex (Dinophyceae). *Journal of Phycology,* 43: 1329–1338.

Margalef, R. (1978). Life forms of phytoplankton as survival alternatives in an unstable environment. *Oceanologica Acta,* 1: 493–509.

McInnery, J.O., Wilkinson, M., Patching, J.W., Embley, T.M., and Powell, R. (1995). Recovery and phylogenetic analysis of novel archaeal rRNA sequences from deep sea deposit feeder. *Applied and Environmental Microbiology,* 61: 1646–1648.

Monaghan, M.T., Balke, M., Gregory, T.R., and Vogler, A.P. (2005). DNA-based species delineation in tropical beetles using mitochondrial and nuclear markers. *Philosophical Transactions of the Royal Society B Biological Sciences,* 360: 1925–1933.

Olmos, J., Paniagua, J., and Contreras, R. (2000). Molecular identification of *Dunaliella* sp. utilizing the 18S rDNA gene. *Letters Applied Microbiology,* 30: 80–84.

Olsen, G.J., Lane, D.J., Ginovannani, S.J., Peace, N.R., and Stahl, D.A. (1986). Microbial ecology and evolution: A ribosomal RNA approach. *Annual Review of Microbiology,* 40: 337–365.

Perumal, P., Sampathkumar, P., and Karuppasamy, P.K. (1999). Studies on bloom farming species of phytoplankton in the Velar Estuary, south east coast of India. *Indian Journal of Marine Science,* 28: 400–401.

Pons, J., Barraclough, T.G., Gomez-Zurita, J., et al. (2006). Sequence-based species delimitation for the DNA taxonomy of undescribed insects. *Systematic Biology,* 55: 595–609.

Rajkumar, R. (2010). Environment Impact Assessment of Mass Cultivation of *Kappaphycus alvarezii* (Doty) Doty ex Silva along the Coast of Palk Bay, Tamil Nadu and the Potential of *Bacillus megaterium* RRM2 Isolated from the Alga for its Proteolytic Activity. Ph.D. dissertation, University of Madras, India.

Redekar, P.D., and Wagh, A.B. (2000a). Growth of fouling diatoms from the Zuari Estuary, Goa (west coast of India) under different salinities in the laboratory. *Seaweed Research and Utilisation,* 22(1&2): 121–124.

Redekar, P.D., and Wagh, A.B. (2000b). Relationship of fouling diatoms number and chlorophyll value from Zuari Estuary, Goa (west coast of India). *Seaweed Research and Utilisation,* 22(1&2): 173–181.

Steele, J.H. (1988). Scale selection for biodynamic theories. In Rothschild, B.J. (Ed.), *Toward a Theory on Biological-Physical Interactions in the World Ocean, NATO ASI Series C: Mathematical and Physical Sciences,* Vol. 239, pp. 513–526. Dordrecht: Kluwer Academic Publishers.

Sullivan, M.J. (1999). Applied diatom studies in estuaries and shallow coastal environments. In *The Diatoms: Application for the Environmental and Earth Sciences.* E.F. Stoermer and J.P. Smol (Eds.), London: Cambridge University Press, pp. 334–351.

Templeton, A.R., Crandall, K.A., and Sing, C.F. (1992). A cladistic analysis of phenotypic associations with haplotypes inferred from restriction endonuclease mapping and sequencing data. III. Cladogram estimation. *Genetics,* 132: 619–633.

UNESCO (United Nations Educational, Scientific and Cultural Organization). (1992). Monitoring the Health of the Oceans: Defining the Role of the Continuous Plankton Recorder in Global Ecosystems Studies. The Intergovernmental Oceanographic Commission and The Sir Alister Hardy Foundation for Ocean Science. IOC/INF-869, SC- 92MS-8.

Van Waasbergen, L.G., Balkwill, D.I., Crockers, F.H., Bjornstad, B.N., and Miller, R.V. (2000). Genetic diversity among *Arthrobacter* species collected across a heterogeneous series of terrestrial deep-subsurface sediments as determined on the basis of 16S rRNA and re cA ngene sequence. *Applied and Environmental Microbiology,* 66: 3454–3463.

Vanormelingen, P., Hegewald, E., Braband, A., Kitschke, M., Friedl, T., Sabbe, K., and Vyverman, W. (2007). The systematics of a small spineless *Desmodesmus* taxon, *D. costatogranulatus* (Sphaeropleales, Chlorophyceae), based on ITS2 rDNA sequence analyses and cell wall morphology. *Journal of Phycology,* 43: 378–396.

Vijayaraghavan, M.R., and Kaur. (1997). *Brown Algae Structure, Ultra Structure and Reproduction.* New Delhi, India: APH Publishing Corporation.

Wiens, J.J. (2007). Species delimitation: new approaches for discovering diversity. *Systematic Biology,* 56: 875–878.

Wiens, J.J., and Penkrot, T.A. (2002). Delimiting species using DNA and morphological variation and discordant species limits in spiny lizards (Sceloporus). *Systematic Biology,* 51: 69–91.

Williams, R. (1993). Evaluation of new techniques for monitoring and assessing the health of large marine ecosystems. In Rapport, D. (Ed.), *NATO Advanced Research Workshop Evaluating and Monitoring the Health of Large-Scale Ecosystems.* Berlin: Springer-Verlag.

Zhang, A.B., Sikes, D.S., Muster, C., and Li, S.Q. (2008). Inferring species membership using DNA sequences with back-propagation neural networks. *Systematic Biology,* 57: 202–215.

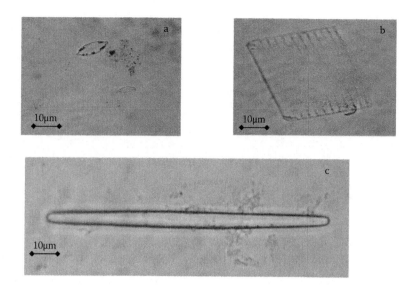

FIGURE 2.1 Morphological diversity of microalgae: (a) *Mastogloia paradoxa* Grun., (b) *Rhabdonema adriatium* Ktz., (c) *Synedra gruvei* Grun.

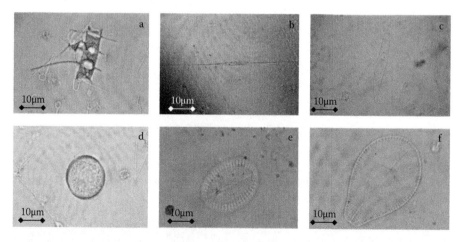

FIGURE 2.2 Morphological diversity of microalgae: (a) *Chaetoceros orientalis* Schiller, (b) *Nitzschia draveillensis* Coste & Ricard, (c) *Pleurosigma formosum* Wm. Smith, (d) *Coscinodiscus janischii* var. *arafurensis* Grun., (e) *Cocconeis scutellum* Ehrenb., (f) *Podocystis spathulata* (Shadbolt) Van Heurck.

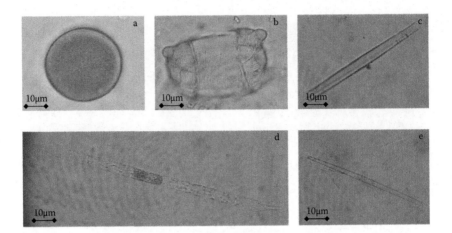

FIGURE 2.3 Morphological diversity of microalgae: (a) *Actinocylus octonarius* Ehrenb., (b) *Biddulphia biddulphiana* (Smith) Boyer, (c) *Thalassionema nitzshioides* Grun., (d) *Rhizosolenia setigera* Btw., (e) *Thalassiothrix longissima* Cleve & Grun.

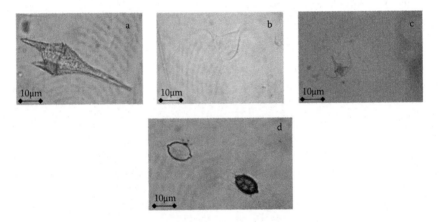

FIGURE 2.4 Morphological diversity of microalgae: (a) *Ceratium hirundinella* (Muller) Dujardin, (b) *Ceratium longipeps* (Bailey) Grun., (c) *Ceratium trichoceros* (Ehrenberg) Kofoid, (d) *Gymnodinium sanguineum* Hirasaka.

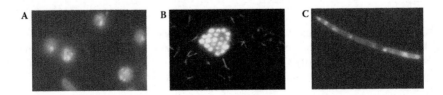

FIGURE 3.2 Nile Red stained *Chlorella* sp.: (a) unidentified chlorophyta, (b) and *Navicula* sp., (c) viewed at 1000× using a Zeiss Axioskop epifluorescence microscope at 490-nm excitation and 585-nm emission filter. Neutral lipid globules in the cytosol are stained yellow. (Unpublished data.)

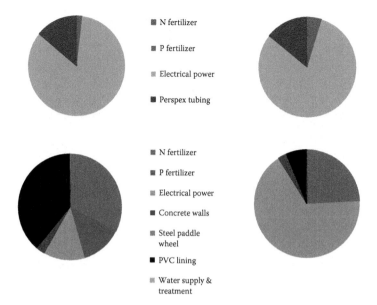

FIGURE 9.2 The relative contribution of fossil energy (left) and GWP (right) to the total requirements for microalgal biodiesel production using a tubular airlift reactor (upper) and a raceway (lower). From the LCA of *C. vulgaris* conducted by Stephenson et al. (2010) under standard conditions. The total fossil energy requirements of 230 and 29 GJ and GWP of 13,550 and 1,900 kg CO_2 per tonne biodiesel formed were estimated for the tubular reactor and raceway, respectively.

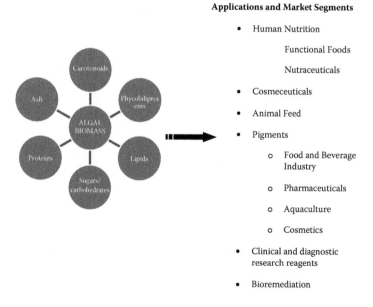

FIGURE 10.1 Applications of algal biomass.

3 Strain Selection for Biodiesel Production

Subburamu Karthikeyan
Department of Agricultural Microbiology
Tamil Nadu Agricultural College
Coimbatore, India

CONTENTS

3.1 BIOPROSPECTING

Bioprospecting is the collection of biological material and the exploitation of its molecular, biochemical, and/or genetic content for the development of a commercial product. Precisely, bioprospecting relies on the endowment of a bioresource, a stock of novel biodiversity. Bioprospecting is a time-consuming process, where new products and markets must be identified, and a compound that covers commercial demands and social needs must be discovered. Algae are ubiquitous and have been evolving as primary biomass producers on the Earth for billions of years. Exploring this existing, self-maintaining, and diverse life form offers a rich base for global biotechnological innovations. Indigenous species are well adapted to prevailing regional abiotic and biotic factors, and further local strains provide an ideal platform for additional strain improvement and process optimization. Many algal species remain unknown or unexplored in science, giving logical attention to explore

TABLE 3.1

Lipid Accumulating Algal Groups in Terms of Abundance

Algae	Representative Lipid Producer	Estimated Number of Described Species	Storage Material	Habitat
Diatoms (Bacillariophyceae)	*Chaetoceros calcitrans, Skeletonoma* sp., *Thalassioria pseudonana, Phaeodactylum tricornutum*	~100,000[d]	Chyrsolaminarin, lipids, polymer of carbohydrates	Oceans, fresh and brackish water, terrestrial
Green algae[a] (Chlorophyceae)	*Botryococcus braunii, Chlorella* spp., *Chlorella vulgaris, Dunaliella salina, Scenedemus* sp., *Ulva* sp.	4,053[e]	Starch and TAGs	Freshwater, terrestrial, marine
Blue-green algae (Cyanophyceae)	*Spirulina* sp.	~2,000[d]	Starch and TAGs	Different habitats
Golden algae (Chrysophyceae)	*Isochrysis* sp.	~1,000[d]	TAGs, leucosin, chrysolaminarin, carbohydrates	Freshwater, marine
Red algae[b] (Rhodophyta)	*Lemanea* fucina, *Gracilaria, Porphyridium cruentum*	6,081[e]	Floridean starch	Mostly marine, freshwater
Brown algae[c] (Phaeophyceae)	*Fucus vesiculosus, Ascophyllum nodosum*	3,067[e]	Laminarin and mannitol	Marine

[a] Seaweeds are included in the green algae (Chlorophyta);
[b] Red algae (Rhodophyta); and
[c] Brown algae (Ochrophyta or Heterokontophyta).
[d] Adapted from Khan et al. (2009).
[e] The World Conservation Union (2010).

this realm for potential application. To further illustrate this point, only fifteen of the currently known microalgal species are mass cultivated in some applied form for use in nutraceuticals, aquaculture feeds, or for wastewater treatment (Raja et al., 2008). Furthermore, the estimated unknown species for all clades of algae are projected to be two orders of magnitude greater than the currently known species (Norton et al., 1996) (Table 3.1). Of the commercialized algae, only a few species are cultivated

TABLE 3.2

Annual Biomass Potential of Microalgae in Comparison to Major Cultivated Crops

Biomass Alga[a]/ Crop[b]	Division	Annual Production	Producer Country	Application
Spirulina	Cyanophyta (cyanobacteria)	3,000 tonnes dry weight	China, India, USA, Myanmar, Japan	Human nutrition Animal nutrition Cosmetics Phycobiliproteins
Chlorella	Chlorophyta (green algae)	2,000 tonnes dry weight	Taiwan, Germany, Japan	Human nutrition Aquaculture Cosmetics
Dunaliella salina	Chlorophyta (green algae)	1,200 tonnes dry weight	Australia, Israel, USA, China	Human nutrition Cosmetics β-Carotene
Aphanizomenon flos-aquae	Cyanophyta (cyanobacteria)	500 tonnes dry weight	USA	Human nutrition
Haematococcus pluvialis	Chlorophyta (green algae)	300 tonnes dry weight	USA, India, Israel	Aquaculture Astaxanthin
Crypthecodinium cohnii	Pyrrophyta (dinoflagellates)	240 tonnes DHA oil	USA	Docosahexaenoic acid (DHA) oil
Schizochytrium spp.	Labyrinthista	10 tonnes DHA oil	USA	Docosahexaenoic acid (DHA) oil
Zea mays (maize)	Magnoliophyta (flowering plants)	868×10^6 tonnes dry weight	Global production	Human nutrition Animal nutrition
Glycine max (soya)	Magnoliophyta (flowering plants)	259×10^6 tonnes dry weight	Global production	Human nutrition Animal nutrition

[a] Adapted from Spolaore et al. (2006).
[b] World Agricultural Supply and Demand Estimates (2012).

at substantial levels, which is trivial when compared to the annual global production of cultivated crops (Table 3.2). To propel algal biotechnological applications to commercially significant sustainable levels, regional species should be investigated for potential application to mass-scale cultivation. The idea of bioprospecting indigenous microalgae for high-value or bioactive products is not innovative. The Aquatic Species Program of National Renewable Energy Laboratory (NREL) stocks more than 3,000 microalgal strains from the United States and Hawaii (Sheehan et al., 1998). Microalgae capable of producing large quantities of docosahexaenoic acid were isolated from marine environments of Western Taiwan (Yang et al., 2010).

Up to now, the key emphasis of microalgal biofuel research has focused on upstream aspects such as bioreactor designs, biomass and lipid production from microalgae, and downstream aspects such as biomass harvesting and the chemistry of oil production.

Microalgal bioprospecting includes isolation of exceptional microalgal strains from aquatic environments for potential value-added products and fine chemicals (Olaizola,

2003; Spolaore et al., 2006). A great deal of literature is accessible on the mass cultivation and sustainable use of microalgae for biofuels; however, relatively few studies have focused on microalgal bioprospecting. Nevertheless, bioprospecting and the establishment of a microalgal collection exclusively for biofuel production have not been reported thus far. Algal bioprospecting or phycoprospecting of indigenous species has an advantage over other methods of sourcing algae from type culture collections and from genetically engineered organisms (Wilkie et al., 2011) (Table 3.3). Screening native algae for species with desirable traits provides a robust biological platform for bioresource production. This biological platform comes equipped with millions of years

TABLE 3.3
Comparison of Different Methods of Sourcing Algae

Method/Source	Merits	Demerits
Phycoprospecting	• Vast diversity of species available • Adapted to local climates and outdoor cultivation • Adapted to local wastewaters and aquatic environments • Adapted to local biota • Native polycultures possible • May provide unique traits amenable to bioresource production • Applicable in any region regardless of access to culture collections • No charge for procurement	• Screening practices must be intensive • Optimization may take dedicated breeding programs • Experiments based on multispecies consortia difficult to translate across laboratories
Culture collections	• Recognized organisms • Unialgal and axenic cultures • Allows comparison between laboratories • Can select for organisms known to produce lipids or high-value compounds • Easy handling • Lower cost of algal inoculant	• Limited number of species available • Unadapted to local climates and outdoor cultivation • May not be able to grow on local wastes • Easily overtaken by native algae in open ponds • May invade local ecosystems
Genetic engineering	• Possibility of increased lipid productivity Production of high-value compounds • May simplify harvesting by excretion of lipids or high-value compounds Modification of traits to increase productivity	• Limited genomic data for algal species • Unadapted to local climates and outdoor cultivation • High cost of development and containment • Negative public perception • Risk of genetic transfer • May invade local ecosystems

Source: Adapted from Wilkie et al. 2011.

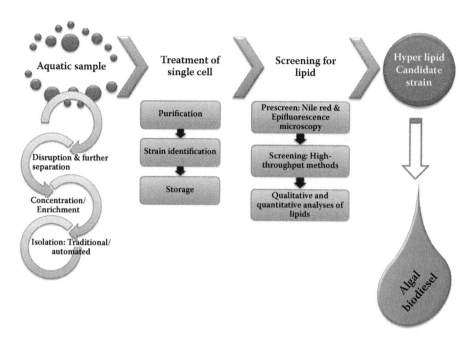

FIGURE 3.1 Schematic outline of procedures in bioprospecting algae for biodiesel production.

of adaptation to the local climate and biota, meaning less energy expended on methods of environmental control and sterile techniques. Specific criteria for the production of biofuels from indigenous algae should include biomass and lipid productivity, harvesting the cells, and oil extractability. Further, the algal oil derived should contain 20%–25% C16 and C18 saturated fatty acid methyl esters and high amounts of unsaturated fatty acid chains, thus offering more cleavage sites to produce hydrocarbons (Gunstone and Harwood, 2007). Phycoprospecting may improve the efficiency of lipid extraction by yielding organisms with traits amenable to oil recovery. For specific objectives such as algal biodiesel, feedstocks for wastewater utilization or mitigation of greenhouse gases (GHGs), the chosen algal strains should satisfy requirements such as the ability to survive in wastewater, capability to grow robustly with higher cell densities, hyper-lipid content as triacylglycerol, and be capable of heterotrophic or mixotrophic growth as wastewater provides both organic and inorganic carbon sources. Until now, research on screening and acclimation of microalgae to adapt to wastewater environments is very sporadic (Zhou et al., 2011). A schematic outline and procedures in bioprospecting algae for biofuel production are outlined in Figure 3.1.

3.2 ISOLATION AND CHARACTERIZATION OF NATURALLY OCCURRING ALGAE

Despite the existence of morphologically diverse algae in a wide variety of terrestrial and aquatic habitats, work with algae has been restricted to a relatively few representatives. This seems partly the result of difficulties encountered in both the isolation and the subsequent purification of the algae. It has been suggested that the techniques normally used to isolate algae may severely limit the number of algal species that can be readily cultured (Castenholz, 1988). The goals of algae isolation and screening efforts are to identify and maintain promising algal specimens for cultivation and strain development. Because it is premature to decide on the system of mass cultivation, new strains should be isolated from a wide variety of environments to provide the largest range of metabolic versatility possible. Algae can be isolated from a variety of natural aquatic habitats, ranging from freshwater to brackish water, to marine and hyper-saline environments, to soil (Round, 1984; Mutanda et al., 2011). Furthermore, large-scale sampling efforts should be coordinated to ensure the broadest coverage of environments and to avoid duplication of efforts. The selection of specific locations can be determined by sophisticated site selection criteria through the combined use of dynamic maps, geographic information systems (GIS) data, and analysis tools. Ecosystems to be sampled could include aquatic (i.e., oceans, lakes, rivers, streams, ponds, and geothermal springs, which includes fresh, brackish, hypersaline, and acidic and alkaline environments) and terrestrial environments in a variety of geographical locations to maximize genetic diversity. Collection sites can include public lands as well as various sites within national and state park systems. In all cases, questions of ownership of isolated strains should be considered. Additionally, within an aqueous habitat, algae are typically found in planktonic (free floating) and benthic (attached) environments. Planktonic algae may be used in suspended mass cultures, whereas benthic algae may find application in biofilm-based production facilities. Sampling strategies should not only account for spatial distribution, but also for the temporal succession brought about by seasonal variations of algae in their habitats (Mutanda et al., 2011).

3.3   ISOLATION TECHNIQUES

For the isolation of new strains from natural habitats, traditional cultivation techniques may be used, such as enrichment cultures (Andersen and Kawachi, 2005). A preferred preliminary step toward single-cell isolations would be enrichment cultures with growth media, soil and/or water extract, or supplementing with nutrients such as nitrate, ammonium, and phosphate or a trace metal. Alternatively, the proximate nutrient composition of the source samples may be analyzed and supplemented to the growth media. Algae survive under natural environments despite the fact that natural samples are often deficient in one or more nutrients. This may be due to the fact that bacterial action, grazing, and death of organisms recycle those nutrients. Sampling reduces the population of specific species that help recycle nutrients, and this nutrient stress leads to a decline of the target species. Enrichment can also be disadvantageous if the target species is unable to compete with the other autochthonous flora. Hence, selective culturing is a unique tool suited for enrichment culturing of lipid-producing microalgae. Once enriched, suspensions containing algae from samples may be centrifuged to increase the biomass concentration of desired cell density; then diluted in sterile water and passed through a 60-μm plankton net to remove zooplankton, and collected again using a 0.45-μm glass filter. The cells on the filter should be rinsed several times with sterile saline to remove bacteria and then inoculate to BG-11 medium (Rippka et al., 1979) for enrichment. However, some algal strains take weeks to months to be isolated by traditional methods (Anderson, 2005). For large-scale sampling and isolation efforts, high-throughput automated isolation techniques involving fluorescence-activated cell sorting (FACS) have proven extremely useful (Sieracki et al., 2005). Because of morphological similarities when comparing many algal species, actual strain identification should be based on molecular methods such as rRNA sequence comparison or, in the case of closely related strains, other gene markers.

3.3.1 MEDIA CONFIGURATION

The distribution of algal species is facilitated by both the selective action of the chemo-physical environment and by the organism's ability to colonize a particular habitat. Numerous culture media have been developed for the isolation and cultivation of microalgae. Some of them are modifications formulated based on the nutrient requirements of the organism. For instance, better growth of marine algae can be achieved by adding small quantities of natural seawater (less than 1% to 4%) rather than supplementing with artificial seawater. Likewise, Schreiber solution, a mixture of nitrate and phosphate, was based on the minimum requirement of the two elements by a diatom culture. Soil extract is amended to Schreiber's medium for cultivating green dasycladalean *Acetabularia* and some unicellular benthic marine algae. Earlier algal media were formulated, to include antibiotics, vitamins, trace metals, and organic chelators such as citrate, which was later replaced with EDTA. Likewise, Chu's medium No. 10 was composed based on proximate analyses of natural samples such as eutrophic lakes. Antibiotics are generally added to the growth medium to inhibit contaminating protists. Germanium dioxide is suggested to inhibit the growth of diatoms. Antibiotics are also helpful as extensive cleansing agents. McDaniel et al. (1962) were able to purify algal cultures free of bacterial contamination using a procedure involving treatment with a detergent and carbolic acid. Various media configured for isolation and cultivation of algae are given in Table 3.4.

3.3.2 TRADITIONAL METHODS

Using a micropipette or Pasteur pipette, or a glass capillary having a straight, bent, or curved tip is handy for single-cell isolation. Micropipettes enable fishing out a single cell from the sample after a series of transfers into sterile rinsing droplets, without the cell being damaged in the process. Finally, the single cell can be pipetted and transferred to the culture medium after microscopic examination. Lewin (1959) recommended placing the droplets on agar to reduce evaporation, but this depends on the size of the cells. Technical skill and expertise are important in order not to shear or damage the cell. The damage may be apparent as cessation of swimming in flagellates or a difference in light refraction due to broken frustules as in diatoms, and severe damage is evident by leakage of protoplasm. The traditional method of micropipette isolation can be successfully attempted with the use of ultra-pure sterile droplets for rinsing, as marine samples hold suspended particles.

TABLE 3.4
Common Media Used for Microalgal Strains from Diverse Aquatic Environments

Media	Freshwater	Marine	Brackish	Suitable for	Ref.
AF6 medium, Modified	+	+	−	Euglenophyceae, volvocalean algae, xanthophytes, many cryptophytes, dinoflagellate and green ciliate; specific for algae requiring slightly acidic medium	Watanabe et al., 2000
AK medium	−	+	+	Broad-spectrum marine algae	Barsanti and Gualtieri, 2006
ASM-1 medium	−	+	+	Marine microalgae	Heaney and Jaworski, 1977
ASN-III medium	−	+	−	Marine Cyanophyceae	Rippka, 1988
ASP –M medium	−	+	ND	Marine macroalgae and microalgae	Goldman and McCarthy, 1978
Beijerinck Medium	+	−	−	Freshwater Chlorophyceae	Andersen et al., 1997
BG-11	+	+	+	Freshwater soil, thermal, and marine Cyanophyceae	Vonshak, 1986
Bold's Basal medium	+	−	−	Broad-spectrum medium for freshwater Chlorophyceae, Xantophyceae, Chrysophyceae, and Cyanophyceae; unsuitable for algae with vitamin requirements	Bold, 1949
C medium	+	−	−	Chlorococcalean algae, some volvocalean algae, some other desmids	Andersen et al., 1997
C30 medium	+	−	−	Freshwater Chlorophyceae	Andersen, 2005
Chu #10 medium	+	−	−	Variety of algae, including green algae, diatoms, cyanobacteria, and glaucophycean alga	Chu, 1942
CHU-11 medium	+	−	ND	Freshwater Cyanophyceae	Nalewajiko et al., 1995
COMBO medium	−	+	+	Cyanobacteria, cryptophytes, green algae, and diatoms	Kilham et al., 1998

TABLE 3.4 (*Continued*)
Common Media Used for Microalgal Strains from Diverse Aquatic Environments

Media	Freshwater	Marine	Brackish	Suitable for	Ref.
Cramer and Myers medium	+	−	−	Euglenophyceae	Nichols, 1973
Diatom medium, modified	+			Freshwater diatom	Cohn et al., 2003
DY V medium	+	+	+	For many algae, especially chlorococcalean algae, filamentous green alga, xanthophycean alga, euglenoid and cyanobacteria	Lehman, 1976
DY V medium	+	−	−	Wide range of heterokont algae, cryptophytes, and other algae that require slightly acidic to circum-neutral pH conditions	Andersen et al., 1997
DYIII medium	+	−	−	Freshwater Chlorophyceae and cyanobacteria	Lehman, 1976
ESAW medium	−	+	ND	Enriched natural seawater medium	Harrison et al., 1980
ESAW medium	−	+	+	Broad spectrum medium for coastal and open ocean algae	Berges et al., 2001
Fraquil medium	+	−	−	For study of trace metal interactions with freshwater phytoplankton	Morel et al., 1975
Guillard's F/$_2$ medium	−	+	+	Broad-spectrum medium for coastal algae; growing coastal marine algae, especially diatoms	Guillard, 1975
Guillard's WC medium	+	−	−	Cyanobacteria, cryptophytes, green algae, and diatoms	Guillard, 1975
Johnson's medium	+	+	+	Broad-spectrum medium	Johnson et al., 1968
K medium	−	+	+	Broad-spectrum medium for oligotrophic marine algae	Andersen, 2005
L1 medium	−	+	+	For oligotrophic (oceanic) marine phytoplankters	Guillard and Hargraves, 1993

(*Continued*)

TABLE 3.4 (*Continued*)
Common Media Used for Microalgal Strains from
Diverse Aquatic Environments

Media	Freshwater	Marine	Brackish	Suitable for	Ref.
MBL medium Woods Hole	+	−	−	Freshwater algae	Nichols, 1973
Medium f	−	+	+	Broad-spectrum medium for marine algae	Jeffrey and LeRoi, 1997
Medium G	+	+	ND	Broad-spectrum medium	Blackburn et al., 2001
MNK medium	−	+	ND	General medium for marine algae, especially coccolithophores	Noel et al., 2004
Sato medium, modified	+	+	+	Freshwater Chlorophyceae	Richmond, 1983
SN medium	−	+	−	Marine cyanophyceae	Waterbury et al., 1986
Walne's medium	−	+	+	Broad-spectrum medium for marine algae (especially designed for mass culture)	Walne, 1970

Source: Adapted from Mutanda et al., 2011.
Note: ND, not determined; +, can be used; −, cannot be used.

Alternatively, many coccoid algae and most soil algae can be isolated on agar plates. It is the preferred isolation method because it is simple and requires no further processing. Streak or pour plating on suitable agar growth medium enables successful isolation, although few algae grow embedded in agar (Brahamsha, 1996). An improvised procedure is to make a fine or atomized spray of cells, usually a liquid cell suspension atomized with sterile air under pressure, which can then be used to inoculate or spread on agar plates. Similarly, a dilution method can be used, wherein a single cell is deposited in a test tube, flask, or well of a multiwell plate (Throndsen, 1978). Selection of the appropriate maximum dilution for plating depends on the probable cell density in natural samples. The dilutions can be effected in several ways, such as dilution with sterile culture medium, distilled water, seawater, and filtered water from the sample site, or some combination of these. Also, where necessary, salts of ammonium, selenium, or another element can be added as supplements to specifically isolate selected species.

When samples contain a wide variety of cells, centrifugation or settling can be foreseen. The target of concentrating the cells instead of obtaining an axenic culture can easily be achieved by gravity. Also, gravity comes in handy when the goal is to separate the larger and heavier cells from smaller algae and bacteria. Specifically for large dinoflagellates and diatoms, moderate centrifugation for a short duration is enough to pelletize them, and smaller cells can be decanted. Density gradient centrifugation with silica sol, Percoll™, etc. has been successfully employed to separate mixed laboratory cultures so that individual species can be separated into a sharp band (Reardon et al., 1979). Large, nonmotile algal cells can be effectively separated by settling. Hence, gravimetric settling is the choice if one aims for concentrating larger cells; however, it is not effective to obtain unialgal culture and hence suggests some combination with other procedures.

3.3.3 ADVANCED METHODS

Although single-celled, colonial, or filamentous algae growing on the agar surface can be isolated by streak plate or spraying, any flagellates as well as other types of algae require the use of advanced techniques. A unialgal culture would contain only one kind of alga, usually a clonal population, but may contain other life forms such as bacteria, fungi, or protozoa. Alternatively, cultures may be axenic, in that they contain only one species of alga. Unialgal cultures are best isolated by targeting the isolation of the zoospores immediately after release from parent cell walls as those cells that begin attaching to surfaces are likely to add contaminants. The algal isolation techniques involving cell separation pose limitations with highly heterogeneous samples or when the cells are suspended in a solution of different chemicals, biomolecules, and cells. This can be overcome by employing a micromanipulator, which successfully permits the separation of a single cell from a liquid culture. The single cell can be easily separated from an enriched environmental sample and grown in liquid medium as monoculture or in agar plates, thus facilitating a significant time saving over the conventional plating technique. The micromanipulator is the ideal tool for algal screening and isolation, provided the person handling it has acquired skill in handling the equipment (Kacka and Donmez, 2008; Moreno-Garrido, 2008). Using micromanipulation techniques requires expertise and skill. It requires the handling of an inverted microscope or stereo zoom microscope with a magnification up to 200×. Phase contrast or dark-field microscopy offers advantages. Capillary tubes or hematocrit tubes of approximately 1 mm diameter × 100 mm long are used for picking individual cells (Godhe et al., 2002; Knuckey et al., 2002).

High-throughput cell sorting is possible when coupled with flow cytometry, which facilitates the rapid and efficient screening of microalgal strains. Microalgae possess different photosynthetic pigments, emitting various auto-fluorescence, which can be applied in flow cytometry to identify algae (Davey and Kell, 1996). Literature on the isolation of microalgae from natural waters employing flow cytometric cell sorting is available (Reckermann, 2000; Crosbie et al., 2003). Chlorophyll is used as a fluorescent probe to distinguish different strains of microalgae. Reckermann (2000) and Sensen et al. (1993) used the chlorophyll auto-fluorescence (CAF) properties of eukaryotic phytoplankton, diatoms, and pico-autotrophic cells for isolation of axenic cultures, whereas Crosbie et al. (2003) used both red and orange auto-fluorescence to differentiate species of algae. Similarly, green auto-fluorescence (GAF), which is common in both autotrophic and heterotrophic dinoflagellates, is also a valuable taxonomic consideration (Tang and Dobbs, 2007). Hence, flow cytometry, coupled with cell sorting, can signify a vital tool for screening and exploiting microalgal strains for specific drives, including biodiesel feedstock development. As compared to fluorescence microscopy, flow cytometry helps the investigator perform rapid and quantitative experimentation. Fluorescence-activated cell sorting (FACS) permits cells with a specific characteristic—or indeed a combination of characteristics—to be separated from the sample. Sinigalliano et al. (2009) compared electronic cell sorting and conventional methods of micropipette cell isolation with dinoflagellates and other marine eukaryotic phytoplankton. Fragile dinoflagellates such as *Karenia brevis* (Dinophyceae) were distressed upon conventional micropipette procedures while cells were viable on electronic sorting. However, electronic single-cell sorting combined with automated techniques for growth screening has the possibility of screening novel algal strains (Sinigalliano et al., 2009). The benefits and shortcomings of the microalgal isolation and purification protocols described in this section are summarized in Table 3.5.

In addition, several immunological and nonimmunological methods to isolate desired unicellular algal cells exist. The immunologic reaction of a specific integrated protein on the membrane decides the protocol for cell separation. Large-scale commercialized cell separation involves techniques such as FACS (Takahashi et al., 2004), magnetic-activated cell sorting (Han and Frazier, 2005), and affinity-based cell sorting (Chang et al., 2005), all of which are highly specific and selective. But the limitation is that the immunologically isolated cells may undergo trauma and the inclusive separation system involves high cost. Further, immunoreactions and follow-up elution with capturing antibodies are quite complicated processes. Alternatively, nonimmunological techniques such as dielectrophoresis (Doh and Cho, 2005), hydrodynamic separation (Shevkoplyas et al., 2005), aqueous two-phase system (Yamada et al., 2002), and ultrasound separation (Petersson et al., 2004) have also been employed. These methods work based on the interactive physico-chemical property of a cell with that of the surrounding media, and lack specificity.

3.4   SCREENING CRITERIA AND METHODS

An ideal screen would consider growth physiology, including cell size and numbers, and metabolite production of algal strains. The algal growth physiology for biofuel encompasses a number of parameters, such as maximum cell density, maximum specific growth rate, and tolerance to environmental variables such as temperature, pH, salinity, oxygen levels, CO_2 levels, and nutrient requirements (Chisti, 2007; Brennan and Owende, 2010). Because all these parameters require significant experimental effort, the development of automated systems that provide information regarding all parameters simultaneously would be helpful. Screening for metabolite production may involve determining the cellular composition of proteins, lipids, and carbohydrates, and measuring the productivity of the organism regarding metabolites useful for biofuel generation. The exact screenings employed would depend on the cultivation approaches and fuel precursor desired. For example, a helpful screening for oil production would allow for distinguishing between neutral and polar lipids, and would provide fatty acid profiles. Furthermore, many strains also secrete metabolites

TABLE 3.5

Advantages and Disadvantages of Microalgal Purification Techniques

Purification Technique	Advantages	Disadvantages	Ref.
Pringsheim's micropipette method	• Single cells can be successively transferred and purified	• Laborious and time-consuming method requiring considerable manual skills • The method often fails with small nonflagellate cells, which are more difficult to recognize during serial transfers • Some delicate flagellates are easily damaged during successive micropipette transfers	Guillard, 1973; Melkonian, 1990
Agar plating (or spraying)	• Relatively easy	• Cannot be used with most flagellate taxa that fail to grow on solid substrates	Hoshaw and Rosowski, 1973
Serial dilution	• Relatively easy	• Unsuccessful when the numerical ratio between algae and bacteria is unfavorable	Brahamsha, 1996
Differential centrifugation	• Less damaging to sensitive cells	• Costly method	Wiedeman et al., 1964
Filtration	• Less damaging to sensitive cells and usually gives better separation of algae from bacteria than differential centrifugation	• It is problematical with small algal cells and with cells secreting mucilage because of bacteria embedded in the mucilage that may also clog filters	Melkonian, 1990
Use of antibiotics	• Relatively easy • Low cost	• Damage the alga as well as leads to increased resistance levels in contaminating bacteria	McDaniel et al., 1962

(Continued)

TABLE 3.5 (*Continued*)
Advantages and Disadvantages of Microalgal Purification Techniques

Purification Technique	Advantages	Disadvantages	Ref.
Flow cytometry	• Precise and rapid method • Simultaneous measurements of individual particle volume, fluorescence and light scatter properties • Highly suitable for separating bacteria from algae to establish axenic algal cultures • Can be used directly in natural samples • Useful for small and delicate taxa	• Requires considerable costs for equipment and its operation • Requires multi-user or central facilities • Axenic cultures are difficult to obtain from algae to which bacteria are physically attached	Sensen et al., 1993
Ultrasonication	• Useful for separating attached bacteria from algal cell walls or mucilage	• Not a stand-alone method • Should be coupled subsequent to cell sorting	Steup and Melkonian, 1981
Immunological methods	• High specificity • Highly selective	• High cost • May cause cell damage	Han and Frazier, 2005; Takahashi et al., 2004

Source: Adapted from Mutanda et al. (2011).

into the growth medium. Some of these could prove valuable as co-products, and new approaches are needed to develop screening methods for extracellular materials. For mass culture of a given algal strain, it is also important to consider the strain's robustness, which includes parameters such as culture consistency, resilience, community stability, and susceptibility to predators present in a given environment. Previous studies revealed that algal strains tested in the laboratory do not always perform similarly in outdoor mass cultures (Sheehan et al., 1998). Therefore, to determine a strain's robustness, small-scale simulations of mass culture conditions must be performed.

At this time, the bottleneck in screening large numbers of algae stems from a lack of high-throughput methodologies that would allow simultaneous screening of multiple phenotypes, such as growth rate and metabolite productivity. Solvent extraction, for example, is the most common method for the determination of lipid content in algae, but it requires a significant quantity of biomass (Bligh and Dyer, 1959; Ahlgren and Merino, 1991). Fluorescent methods using lipid-soluble dyes have also been described, and although these methods require much less biomass (as little as a single cell), it has not yet been established if these methods are valid across a wide range of algal strains (De la Jara et al., 2003; Elsey et al., 2007). Further improvements in analytical methodology could be made through the development of solid-state screening methods. Not only are rapid screening procedures necessary for the biofuels field, but they also could prove extremely useful for the identification of species, particularly in mixed field samples necessary for the future of algal ecology. They could also reduce the number of redundant screens of algal species.

3.5   SCREENING AND SELECTION FOR LIPID PRODUCTION

Conventional methods of solvent extraction and gravimetric determination for lipid quantification (Bligh and Dyer, 1959) are laborious and time consuming. Moreover, approximately 10 to 15 mg wet weight of cells (Akoto et al., 2005) must be cultured for any appreciable extraction and derivatization. However, in-situ lipid content measurements would significantly reduce the quantity of sample as well the preparation time required. Accordingly, there is greater interest in a rapid in-situ measurement of the lipid content of algal cells (Cooksey et al., 1987). Nile Red (9-diethylamino-5H-benzo[α]phenoxazine-5-one), a lipid-soluble fluorescent dye, has been commonly used to evaluate the lipid content of animal cells and microorganisms such as yeasts and fungi (Genicot et al., 2005) and specifically extended to microalgae (Cooksey et al., 1987; Elsey et al., 2007). Nile Red is relatively photostable and produces intense fluorescence in organic solvents and hydrophobic environments, which makes them a better candidate for in-situ screening for lipids. Furthermore, neutral and polar lipids can be clearly differentiated due to polarity changes in the medium as evinced by a blue shift in the emission maximum of Nile Red (Greenspan and Fowler, 1985; Laughton, 1986; Cooksey et al., 1987; Lee et al., 1998). The solvent system used for Nile Red would determine the emission spectra of the dye (Elsey et al., 2007). However, the thick cell walls of microalgae inhibit the permeation of Nile Red, and this is variable among algal species, requiring the use of high levels of solvents such as DMSO (20% to 30% v/v) and elevated temperatures (40°C) (Chen et al., 2009). Then again, Chen et al. (2011) developed a two-step microwave-assisted staining method for in vivo quantification of neutral lipids in green algae with thick, rigid cell walls that prevents penetration of the Nile Red dye into the cell. This may also be appropriate for other classes of algae that do not stain properly with Nile Red. Hence, a Nile Red assay can be used as a tool for screening oleaginous algal strains as well as quantitatively determining the neutral lipids in algal cells (Figure 3.2; see color insert).

Recently, another class of lipophilic fluorescent dye BODIPY® 505/515 (4,4-difluoro-1,3,5,7-tetramethyl-4-bora-3a,4a-diaza-s-indacene) has been used to potentially stain microalgal lipids. BODIPY staining lets the lipid droplets stain green and the chloroplasts stain red in live algal cells (Cooper et al., 2010). BODIPY 505/515 is advantageous over Nile Red in emitting a narrower spectrum (Cooper et al., 2010;

 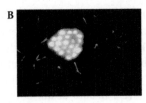 

FIGURE 3.2 (See color insert.)   Nile Red stained *Chlorella* sp.: (a) unidentified chlorophyta, (b) and *Navicula* sp., (c) viewed at 1000× using a Zeiss Axioskop epifluorescence microscope at 490-nm excitation and 585-nm emission filter. Neutral lipid globules in the cytosol are stained yellow. (Unpublished data.)

Govender et al., 2012). This facilitates the fluorescence distinction of lipid bodies, resulting in better resolution and thus is important for seamless confocal imaging (Cooper et al., 1999). Furthermore, unlike Nile Red, BODIPY 505/515 does not fix to cytoplasmic constituents other than lipid bodies and chloroplasts. This discerning property of BODIPY 505/515 to bind to lipid bodies alone offers rapid screening and isolation of hyper-lipid producing algal strains. Bigelow et al. (2011) developed a rapid, single-step, laboratory-scale in-situ protocol for GC–MS (gas chromatography with mass spectroscopy) lipid analysis that requires only 250 µg dry mass per sample. When coupled with fluorescent techniques using Nile Red or BODIPY dyes and flow cytometry for cell sorting, the aforesaid GC-MS analysis allows throughput screening of lipid-producing algal strains from varied environments. Upon isolation, purification, and identification of a hyper-lipid producing algal strain, the researcher would be interested in the physiological traits such as the photosynthetic efficiency, carbon fixation rate, growth rate, etc. Alternatively, infrared analysis, which does not depend on stain application but rather detects specific molecular absorption bands to give approximate concentrations, can be used for the detection of many metabolites, including lipids. This method has recently been applied to detecting changes in algal cell composition during nitrogen starvation (Dean et al., 2010).

Spectroscopic methods such as near-infrared (NIR) and Fourier transform infrared (FTIR) spectroscopies have been established to predict the levels of spiked polar and neutral lipids in algal cells based on multivariate calibration models (Laurens and Wolfrum, 2011). The above infrared spectroscopic techniques are rapid, high-throughput, and non-destructive means of algal screening for lipids. Hence, this calibration model serves as a short-time, high-throughput method of quantifying cell lipids compared to time-consuming traditional wet chemical methods. The NIR and FTIR spectra of biomass of various species accurately predicted the levels of lipids. This fast, high-throughput spectroscopic lipid fingerprinting method is pragmatic in real-time monitoring of lipid accumulation or a multitude of screening efforts that are ongoing in the microalgal research community. Coherent anti-Stokes Raman scattering microscopy is also an associated technique that creates an image of whole cells based on the vibrational spectra of a specific cellular constituent. Huang et al. (2010) demonstrated that Stokes Raman spectroscopy could accomplish detection and identification of cellular storage lipids, specifically triglycerides. Further, similar to infrared spectroscopic techniques, Raman scattering microscopy is also prospective as a rapid, noninvasive compositional analysis method that enables imminent in-line or at-line lipid monitoring. Recently, a single-cell, laser-trapping Raman spectroscopic method that is direct and in vivo has been described as an efficient tool for profiling microbial cellular lipids (Wu et al., 2011). This method is proven in the quantitative estimation of the degree of unsaturation and transition temperatures of algal cellular lipids.

3.6 PRESERVATION

3.6.1 TRANSFER TECHNIQUES

The accomplishment of bioprospecting rests with the successful long-term maintenance of the algal strains. The most common method used to preserve microalgal cultures is perpetual maintenance under a controlled environment. Periodic serial sub-culturing of the mother culture onto agar slants is done to maintain the strains (Day, 1999; Warren et al., 2002; Richmond, 2004). This provides metabolically active cultures that retain a vigorous, morphologically, physiologically, and genetically representative population. A crucial factor to consider is that different durations of sub-cultures may provide different stages of the life cycle. Proper labeling and careful checking are required before starting a serial transfer. Manipulations and transfer should be carried out under aseptic conditions. Rigorous microbiological methods, following standard guidelines for aseptic techniques and maintenance procedures, are crucial (Isaac and Jennings, 1995). The revival of preserved cultures can be successfully accomplished with 1% to 10% (v/v) of the original culture, but some dinoflagellates, *Synechococcus* and *Prochlorococcus*, may require a higher inocula level of up to 25% (v/v). Another issue with agar cultures is that some benthic diatom colonies may stick rather firmly to the agar surface or that some filamentous cyanobacteria may even grow into the agar. These can be transferred by removal of agar along with algal material. If older agar cultures are to be revived, the slant may be over-layered with fresh liquid medium for several hours prior to transfer. Another issue that needs bearing in mind is that not all species form appreciable colonies on agar, specifically many flagellates and other planktonic species; likewise, few edaphic and aquatic benthic microalgae grow well in liquid medium. Regardless of that, slant cultures are preferred because of easy and minimal handling during transfer and hence a lower risk of contamination. Mixing the algal liquid culture is customary during transfer to fresh medium or plates. However, uncontrolled mixing bears the risk of damage to delicate coccoid green algae, cyanobacteria, and some fragile diatoms such as *Thalassiosira* and *Rhizosolenia*. But mixing may be mandatory in some instances or, as in case of *Polytoma,* where resting cells settle to the bottom, the cell transfer should effect from the bottom of the culture vessel (Anderson, 2005). In contrast, certain colonial flagellate and coccoid green algae (*Eudorina, Pediastrum*) need agitation and aeration to obtain typical morphology.

3.6.2 MAINTENANCE CONDITIONS

The maintenance of metabolically active algae is essential because of the conservation of stock cultures, attainment of explicit morphological and physiological status, or mass production. As described earlier, conservation of stock cultures is by routine, serial sub-culture and storage, preferably under suboptimal temperature and light regimes that may be similar for most algae. In addition, the nature of the media also plays a role in the frequency of transfer interval. But, for the achievement of desirable physiological cultures or for mass cultures, optimal growth conditions are vital, and this varies greatly with strains. In fact, algae poorly adapted to a specific medium may alter morphological features, as in the case of *Chlamydomonas*, where loss of functional flagella and some cyanobacteria may lose cell surface features. Another concern needing emphasis, specifically in continuous culture systems, is that culture conditions such as pH, nutrient content, oxygen level, etc. tend to change over time, despite the fact that the external environment remains unchanged and the limiting substrate concentration is at the required concentration. Some microalgae having an absolute requirement for vitamin B12 at very low concentrations can be grown without supplementing vitamin B12 in the culture medium for a number of generations. Complementing medium with vitamins B1 or B12 helps in stimulating the growth of certain algae. Another intrinsic phenomenon of some diatoms is that the cell size eventually becomes too small during continuous vegetative propagation to remain viable. A better alternative is to allow sexual reproduction of the culture to regenerate large, new vigorous cells. To propagate indefinitely, some Dasycladales are subjected to undergo periodic sexual reproduction.

One should appraise whether a particular alga strain would be best maintained for long periods in liquid medium or in agar slants. This is influenced by many environmental factors, including the habitat of the strain. A soil–water biphasic medium favors the growth of filamentous green algae and euglenoids. In fact, the addition of a soil phase directs the coccoid green algae to retain morphology, and a medium without soil extract promotes the accumulation of starch granules or lipid droplets. Hence, a choice of suitable culture medium specific to the strain is crucial. Second, light intensity must also be considered. For long-term culturing and maintenance of most microalgae, coupling subdued temperature with light intensities between 10 and 30 mmol photons $m^{-2}s^{-1}$ is vital. Excessive light can cause photo-oxidative stress in some algae. That is one of the reasons that some marine algae of tropical open-ocean are killed by continuous light (Graham and Wilcox, 2000). Furthermore, low light intensities are usually preferred by algae with phycobilisomes, while most dinoflagellates often need higher light intensities (60 to 100 mmol photons $m^{-2}s^{-1}$). This directs most culture collections to vary the light:dark regimens between 12:12 and 16:8 hours light:dark. However, preserving algae from extreme environments needs specific insight, as suggested by Elster et al. (2001). Third, the temperature of storage is vital. Variations in temperature can more easily influence marine strains than freshwater strains. In general, microalgal cultures are successfully conserved at temperatures between 15°C and 20°C. Indeed, some larger service repositories such as the Culture Collection of Algae at the University of Texas (UTEX) preserve algal strains at 20°C. Prolonged maintenance at 20°C leads to cellular damage resulting from photo-inhibition. And, alternatively, increased light intensities coupled with incubation temperatures higher than 20°C can be employed. However, temperatures above 20°C are mostly incorrect for conserving stocks at comprehensive transfer cycles. One should note that the evaporation of the medium effectively regulates the interval of their transfer cycles. Fourth, the frequency of transfer cycles is considered a key factor. For routine maintenance, sub-culturing is done toward the end of the exponential growth phase of the culture. The shortest transfer cycle is about 1 to 2 weeks for sensitive strains, while some green algae and cyanobacteria, on agar slants at low light and 10°C, is sub-cultured only once every 6 months. However, a safe transfer interval for a specific strain can be predictable to one quarter the time a strain can survive maximally. Usually, a post-transfer period at higher light and temperature regime is valuable in regular quality control assessment. Moreover, perpetual maintenance over longer periods may alter the morphological features and physiological characteristics of some strains, and hence a short interval of maintenance at optimal growth conditions is recommended to refresh the culture. In case of an unknown or a newly isolated strain, the cultural characteristics should be fully understood prior to maintenance, or the long-term maintenance procedure should be framed by optimization of survival at varied light and temperature conditions (Lokhorst, 2003). It is often sensible to screen and configure a suitable medium for the new isolate. Finally, the culture maintenance chamber or room should be controlled for humidity—for not only preventing evaporation of cultures, but also to avoid the contaminating fungi and molds.

In spite of these, the selective and synthetic nature of the media as well the incubation conditions, as opposed to the native ecological conditions, limit the success of perpetual transfer. Furthermore, continuous transfers lead to the loss of morphological and/or genetic characteristics (Warren et al., 2002). Not to mention that serial sub-culturing is a labor-intensive and time-consuming process, which restricts maintenance and handling of a large number of cultures. Above all, to overcome the risk of loss of strains, each strain may be maintained in secondary culture collections with sub-cultures of different ages or transfer dates. The World Federation of Culture Collections has suggested to stock backup cultures at various locations to expedite all possible chances of revival (Anon, 1999).

3.6.3 CRYOPRESERVATION

Perpetual transfer leading to long-term culturing, usually under conditions very different from its natural environment, leads to genetic variants among the population adapted to the artificial culturing environment. To sidestep the shortcomings of serial sub-culturing, alternate methods of ex-situ conservation of algal strains are suggested. Continuous maintenance of actively growing algal strains on a long-term basis is often costly, and time and labor consuming. In contrast, cultures can be maintained alive in a retarded metabolic state that requires less attention. One approach is to maintain resting spores or other dormant stages of some algal species (such as akinetes) at ambient temperatures for many years without any attention. Leeson et al. (1984) were able to recover aplanospores of *Haematococcus pluvialis* Flotow from air-dried soil even after 27 years. However, it should be considered that the viability of resting stages generally declines with time, and many aquatic algae do not show any insistent dormant stage. Hence, the addition of bacteriostatic chemicals and agents that prevent autolysis of algal cells to help improve the cell viability during the entire storage time is generally recommended. Some of the major preservatives in use today are formalin, 1% Lugol solution, and 3% glutaraldehyde (Wetzel and Likens, 2000). The concentration can be altered based on the type and nature of the algae and maintenance conditions. For instance, a 3% glutaraldehyde concentration is too high, causing withering or complete disintegration of cells beyond the ability for retention of normal cell shape, specifically of wall-less flagellates.

To overcome the shortcomings and inclusion of chemicals in maintenance medium, lyophilization has been accepted widely as a means of conserving viable cultures of all microorganisms in a desiccated state. However, lyophilization involves vacuum desiccation under freezing and subsequent thawing, so cell revival mandates inclusion of cryoprotective agents at high concentrations to offer protection from damage. The cryoprotectants extensively used for algae are methanol, dimethylsulfoxide (DMSO), and glycerol (Taylor and Fletcher, 1998). Methanol and DMSO are preferred for freshwater and terrestrial microalgal cryopreservation, while glycerol and DMSO are useful for marine phytoplanktons (Day et al., 2000). The above are penetrating cryoprotectants and passively move through the plasma membrane to equilibrate between the cell interior and the extracellular solution. Penetrating cryoprotectants are toxic at high concentrations (Adam et al., 1995; Santarius, 1996). Hence, permeating cryoprotectants should be added prior to cryopreservation and should immediately be removed after thawing. Algal spore preservation is heavily dependent on bacterial contamination. Hence, preservation of spores of the green seaweeds *Ulva fasciata* and *U. pertusa* was improved by the addition of ampicillin in f/2 medium at 4°C (Bhattarai et al., 2007).

Cryopreservation is most suited for algae that do not require that the normal resting stage be maintained indefinitely. Because microalgae are cryopreserved as large populations of algal units, the percentage of viability of identical cultures is of great concern and often varies. However, with proper physiological conditioning prior to freezing, the variability can be minimized. This is one of the key reasons that, to date, most dinoflagellates, cryptophytes, synurophytes, and raphidophytes are not successfully cryopreserved. In contrast, most marine diatoms can be effectively cryopreserved, with high viability, although freshwater diatoms fail to revive and have thus proven more problematic. Examinations of large numbers of strains have taken place at the four major protistan collections: Culture Collection of Algae and Protozoa (CCAP) (United Kingdom), The Provasoli-Guillard National Center for Culture of Marine Phytoplankton (CCMP) (United States), Sammlung von Algenku Huren Göttingen (SAG) (Germany), and The Culture Collection of Algae at UTEX (United States); examination reveals that chlorarachniophytes, eustigmatophytes, pelagophytes, phaeothamniophytes, and ulvophytes also have very high success rates, comparable with the other green algae and cyanobacteria. Algal strains that have been reestablished at NREL are being cryopreserved in an effort to reduce the workload associated with maintaining an algae collection and to prevent unintended loss or genetic drift, a risk associated with frequent transfer. The cryofreezer uses liquid nitrogen, and cultures are stored at −195°C in the vapor phase. Nevertheless, it has been distinguished that virtually all large cell sized algae, and most filamentous forms, cannot as yet be cryopreserved. Attempts to determine the fundamental reasons for this failure of cryopreservation on large and complex algae are not satisfying. This warrants auxiliary research on the basic mechanisms of freezing damage. Furthermore, the pragmatic development of improved techniques will expand the number and diversity of algal taxa that can be successfully cryopreserved.

3.7 ROLE OF REPOSITORIES

Repositories are indispensable in preserving the diversity of natural habitats, protecting genetic material, and providing basic resources for research. At present, only a few major algal collection centers exist in the United States and other countries. They currently maintain thousands of different algal strains and support the research and industrial community with their expertise in algae biology. The function of a culture collection often transcends simple depository functions. They may also support research on determining strain characteristics, cryopreservation, and phylogeny, either by themselves or in connection with outside collaborators. Currently, no central database exists that provides global information on the characteristics of currently available algal strains. Protection of intellectual property in private industry has further exacerbated the flow of relevant strain data. Some minimal growth information is available from existing culture collections, but it is very difficult to obtain more detailed information on growth, metabolites, and the robustness of particular existing strains. The establishment of a central strain, open-access repository could accelerate research on algae-based biofuel production systems.

Above all, it is certain that many algal strains in established collections have been cultivated for several decades, and some may have lost their original properties, such as mating capability or versatility regarding nutrient requirements. To obtain versatile and robust strains that can be used for mass culture in biofuel applications, it would be prudent to consider the isolation of new native strains directly from unique environments. For both direct breeding and metabolic engineering approaches to improve biofuels production, it will be important to isolate a wide variety of algae for assembly into a culture collection that will serve as a bioresource for further algal biofuel research.

3.8 CONCLUDING REMARKS

The detection of new and rare species is made easier due to the accessibility of classifications based on genotypic and phenotypic data. This will be valuable in the challenges facing systematic classification and the need for establishing well-defined taxa, a stable nomenclature, and enhanced identification procedures. Large-scale screening for bioactive compounds of industrial application necessitates rapid and unequivocal characterization of enormous numbers of algal isolates. Because these biocatalytic compounds hold persistent value as an input for the biotechnology industry, the conservation of microbial gene pools is critical. Ex-situ collections are and will continue to be an essential cradle for warranting that a source of living cells is available for research and manufacturing purposes. It is well documented that exploring the same or similar environments fails to reveal the same organisms again or even, if found, they would not exhibit the desired characteristics exhibited by the earlier strains. Nevertheless, maintenance of representatives of all identified species of algae and cell lines in ex-situ collections is unrealistic. Hence, it is suggested that future researchers and repositories should ensure the provision of the DNA rather than the organisms themselves. We are still largely in the hunter-and-gatherer stage of exploiting algae for food, bioactive compounds, and energy. Hence, further challenges in bioprospecting may includee the protection of intellectual property rights of original owners, a policy for strain distribution, and sharing and material transfer agreements.

REFERENCES

Adam, M.M., Rana, K.J., and McAndrew, B.J. (1995). Effect of cryoprotectants on activity of selected enzymes in fish embryos. *Cryobiology,* 32: 92–104.

Ahlgren, G., and Merino, L. (1991). Lipid analysis of freshwater microalgae: A method study. *Arch. Hydrobiol.,* 121: 295–306.

Akoto, L., Pel, R., Irth, H., Brinkman, U.A.T., and Vreuls, R.J.J. (2005). Automated GC–MS analysis of raw biological samples: Application to fatty acid profiling of aquatic micro-organisms. *J. Anal. Appl. Pyrol.,* 73: 69–75.

Andersen, R.A. (2005). *Algal Culturing Techniques.* Elsevier Academic Press, Burlington, MA.

Andersen, R.A., and Kawachi, M. (2005). Traditional microalgae isolation techniques. In R.A. Anderson (Ed.), *Algal Culturing Techniques,* Elsevier Academic Press, Burlington, MA, pp. 83–100.

Andersen, R., Morton, S.L., and Sexton, J.P. (1997). CCMP—Provasoli–Guillard National Center for Culture of Marine Phytoplankton 1997 list of strains. *J. Phycol.,* 33 (suppl): 1–75.

Anon. (1999). World Federation for Culture Collections: Guidelines for the Establishment and Operation of Collections of Cultures of Microorganisms. Michael Grunenberg GmbH, Schoeppenstedt, Germany, p. 24.

Barsanti, L., and Gualtieri, P. (2006). *Algae—Anatomy, Biochemistry, and Biotechnology.* Boca Raton, Florida: CRC Press, pp. 215–235.

Berges, J. A., Franklin, D. J., and Harrison, P. J. (2001). Evolution of an artificial seawater medium: Improvements in enriched seawater, artificial water over the past two decades. *J. Phycol.,* 37: 1138–1145.

Bhattarai, H., Paudel, B., Hong, Y.K., and Shin, H. (2007). A simple method to preserve algal spores of *Ulva* spp. in cold storage with ampicillin. *Hydrobiologia,* 592(1): 399.

Bigelow, N.W., Hardin, W.R., Barker, J.P., Ryken, S.A., MacRae, A.C., and Cattolico, R.A. (2011). A comprehensive GC–MS sub-microscale assay for fatty acids and its applications. *J. Am. Oil Chem. Soc.,* 88(9): 1329–1338.

Blackburn, S.I., Bolch, C.J.S., Haskard, K.A., and Hallegraeff, G.M. (2001). Reproductive compatibility among four global populations of the toxic dinoflagellate *Gymnodinium catenatum* (Dinophyceae). *Phycologia,* 40(1): 78–87.

Bligh, E.G., and Dyer, W.J. (1959). A rapid method of total lipid extraction and purification. *Can. J. Biochem. Physiol.,* 37: 911–915.

Bold, H.C. (1949). The morphology of *Chlamydomonas chlamydogama* sp. nov. *Bull. Torrey Bot. Club.,* 76: 101–108.

Brahamsha, B. (1996). A genetic manipulation system for oceanic cyanobacteria of the genus *Synechococcus. Appl. Environ. Microbiol.,* 62: 1747–1751.

Brennan, L., and Owende, P. (2010). Biofuels from microalgae – A review of technologies for production, processing, and extractions of biofuels and co-products. *Renewable and Sustainable Energy Reviews,* 14: 557–577.

Castenholz, R.W. (1988). Culturing methods for cyanobacteria. *Methods Enzymol.,* 167: 68–93.

Chang, W.C., Lee, L.P., and Liepmann, D. (2005). Biomimetic technique for adhesion-based collection and separation of cells in a microfluidic channel, *Lab Chip,* 5: 64–73.

Chen, W., Sommerfeld, M., and Hu, Q. (2011). Microwave-assisted Nile Red method for in vivo quantification of neutral lipids in microalgae. *Bioresource Technol.*, 102(1): 135–141. Available at <http://www.ncbi.nlm.nih.gov/pubmed/20638272>.

Chen, W., Zhang, C., Song, L., Sommerfield, M., and Hu, Q. (2009). A high throughput Nile Red method for quantitative measurement of neutral lipids in microalgae. *J. Microbiol. Methods*, 77(1): 41–47.

Chisti, Y. (2007). Biodiesel from microalgae. *Biotechnol. Adv.*, 25: 294–306.

Chu, S.P. (1942). The influence of the mineral composition of the medium on the growth of planktonic algae. Part I. Methods and culture media. *J. Ecol.*, 30: 284–325.

Cohn, S.A., Farrell, J.F., Munro, J.D., Ragland, R.L.,Weitzell, R.E., Jr., and Wibisono, B.L. (2003). The effect of temperature and mixed species composition on diatom motility and adhesion. *Diatom Res.*, 18: 225–243.

Cooksey, K.E., Guckert, J.B., Williams, S.A., and Callis, P.R. (1987). Fluorometric determination of the neutral lipid-content of microalgal cells using Nile Red. *J. Microbiol. Methods*, 6(6): 333–345.

Cooper, M.S., D'Amico, L.A., and Henry, C.A. (1999). Confocal microscopic analysis of morphogenetic movements. *Methods Cell Biol.*, 59: 179–204.

Cooper, M.S., Hardin, W.R., Petersen, T.W., and Cattolico, R.N. (2010). Visualizing green oil in live algal cells. *J. Biosci. Bioeng.*, 109: 198–201.

Crosbie, N.D., Pockl, M., and Weisse, T. (2003). Rapid establishment of clonal isolates of freshwater autotrophic picoplankton by single-cell and single-colony sorting. *J. Microbiol. Methods*, 55: 361–370.

Davey, H.M., and Kell, D.B. (1996). Flow cytometry and cell sorting of heterogeneous microbial populations: The importance of single-cell analyses. *Microbiol. Rev.*, 60: 641–696.

Day, J.G. (1999). Conservation strategies for algae. In Benson, E.E. (Ed.), *Plant Conservation Biotechnology*. Taylor and Francis Ltd., London, pp. 111–124.

Day, J.G., and Brand, J.J. (2005). Cryopreservation methods for maintaining microalgal cultures. In Andersen, R.A. (Ed.), *Algal Culturing Techniques*. Elsevier Academic Press, Burlington, MA.

Day, J.G., Fleck, R.A., and Benson, E. E. (2000). Cryopreservation recalcitrance in microalgae: Novel approaches to identify and avoid cryo-injury. *J. Appl. Phycol.*, 12: 369–377.

De la Jara, A., Mendoza, H., Martel, A., Molina, C., Nordstron, L., de la Rosa, V., and Diaz, R. (2003). Flow cytometric determination of lipid content in a marine dinoflagellate, *Crypthecodinium cohnii*. *J. Appl. Phycol.*, 15: 433–438.

Dean, A.P., Sigee, D.C., Estrada, B., and Pittmann, J.K. (2010). Using FTIR spectroscopy for rapid determination of lipid accumulation in response to nitrogen limitation in freshwater microalgae. *Bioresour. Technol.*, 101: 4499–4507.

Doh, I., and Cho, Y.H. (2005). A continuous cell separation chip using hydrodynamic dielectrophoresis (DEP) process. *Sensor Actuat. A-Phys.*, 12: 59–65.

Elsey, D., Jameson, D., Raleigh, B., and Cooney, M.J. (2007). Fluorescent measurement of microalgal neutral lipids. *J. Microbiolog. Methods*, 68: 639–642.

Elster, J., Seckbach, J., Vincent, W.F., and Lhotsky, O. (Eds.) (2001). *Algae and Extreme Environments. Ecology and Physiology. Nova Hedwigia*, Suppl. 123, p. 602.

Genicot, G., Leroy, J.L.M.R., and Van Soom, A. (2005). The use of a fluorescent dye, Nile Red, to evaluate the lipid content of single mammalian oocytes. *Theriogenology*, 63: 1181–1194.

Godhe, A., Anderson, D.M., and Rehnstam-Holm, A.S. (2002). PCR amplification of microalgal DNA for sequencing and species identification: Studies on fixatives and algal growth stages. *Harmful Algae*, 27: 1–8.

Goldman, J.C., and McCarthy, J.J. (1978). Steady state growth and ammonium uptake of a fast growing marine diatom. *Limnol. Oceanogr.*, 23: 695–703.

Govender, T., Ramanna, L., Rawat, I., and Bux, F. (2012). Bodipy, an alternative to Nile Red staining technique for intracellular lipid evaluation, *Bioresour. Technol.,* 114: 507–511.

Graham, L.E., and Wilcox, L.W. (2000). *Algae.* Prentice Hall, Upper Saddle River, NJ, p. 700.

Greenspan, P., and Fowler, S.D. (1985). Spectrofluorometric studies of the lipid probe, Nile Red. *J. Lipid Res.,* 26: 781–788.

Guillard, R.R.L. (1973). Methods for microflagellates and nannoplankton. In *Handbook of Phycological Methods. Culture Methods and Growth Measurements*, J. Stein (Ed.), 69–85. New York: Cambridge University Press.

Guillard, R.R.L. (1975). Culture of phytoplankton for feeding marine invertebrates. In W.L. Smith and M.H. Chantey (Eds.), *Culture of Marine Invertebrate Animals.* New York: Plenum Publishers, pp. 29–60.

Guillard, R.R.L., and Hargraves, P.E. (1993). *Stichochrysis immobilisis* a diatom, not a chryso-phyte. *Phycologia,* 32: 234–236.

Gunstone, F.D., and Harwood, J.L. (2007). Occurrence and characterisation of oils and fats. In Gunstone, F.D., Harwood, J.L., and Dijkstra, J.L., Eds., *The Lipid Handbook.* CRC Press; Boca Raton, FL, pp. 37–141.

Han, K.H., and Frazier, A.B. (2005). A microfluidic system for continuous magnetophoretic separation of suspended cells using their native magnetic properties. *Proc. Nanotech.,* 1: 187–190.

Harrison, P.J., Waters, R.E., and Taylor, F.J.R. (1980). A broad spectrum artificial seawater medium for coastal and open ocean phytoplankton. *J. Phycol.,* 16: 28–35.

Heaney, S.I. and Jaworski, G.H.M. (1977). A simple separation technique for purifying micro-algae. *Br. Phycol. J.,* 12: 171–174.

Hoshaw, R.W., and Rosowksi, J.R. (1973). Methods for microscopic algae. In *Handbook of Phycological Methods. Culture Methods and Growth Measurements*, J. Stein (Ed.), 53–67. New York: Cambridge University Press.

Huang, Y.Y., Beal, C.M., Cai, W.W., Ruoff, R.S., and Terentjev, E.M. (2010). Micro-Raman spectroscopy of algae: Composition analysis and fluorescence background behavior. *Biotechnol. Bioeng.,* 105: 889–898.

Isaac, S., and Jennings, D. (1995). *Microbial Culture.* Scientific Publishers Ltd., Oxford, UK, pp. 115–121.

Jeffrey, S.W., and LeRoi, J.-M. (1997). Simple procedures for growing SCOR reference micro-algal cultures. In S.W. Jeffrey, R.F.C. Mantoura, and S.W. Wright (Eds.) *Phytoplankton pigments in oceanography: Monographs on oceanographic methodology* 10, France: UNESCO, pp. 181–205.

Johnson, M.K., Johnson, E.J., MacElroy, R.D., Speer, H.L., and Bruff, B.S. (1968). Effects of salts on the halophylic alga *Dunaliella viridis. J. Bacteriol.,* 95, 1461–1468.

Kacka, A., and Donmez, G. (2008). Isolation of *Dunaliella* spp. from a hypersaline lake and their ability to accumulate glycerol'. *Bioresour. Technol.,* 99: 8348–8352.

Khan, S.A., Rashmi, Hussain, M.Z., Prasad, S., and Banerjee, U.C. (2009). Prospects of bio-diesel production from microalgae in India. *Renewable and Sustainable Energy Rev.,* 13: 2361–2372.

Kilham, S.S., Kreeger, D.A., Lynn, S.G., Goulden, C.E., and Herrera, L. (1998). COMBO: A defined fresh water culture medium for algae and zooplankton. *Hydrobiologia,* 377: 147–159.

Knuckey, R.M., Brown, M.R., Barrett, S.M., and Hallegraeff, G.M. (2002). Isolation of new nanoplanktonic diatom strains and their evaluation as diets for juvenile Pacific oysters (*Crassostrea gigas*). *Aquaculture,* 211: 253–274.

Laughton, C. (1986). Measurement of the specific lipid content of attached cells in microtitre cultures. *Anal. Biochem.,* 156: 307–314.

Laurens, L.M.L., and Wolfrum, E.J. (2011). Feasibility of spectroscopic characterization of algal lipids: Chemometric correlation of NIR and FTIR spectra with exogenous lipids in algal biomass. *Bioenergy Res.*, 4: 22–35.

Lee, S.J., Yoon, B.D., and Oh, H.M. (1998). Rapid method for the determination of lipid from the green alga *Botryococcus braunii*. *Biotechnol. Tech.*, 12: 553–556.

Leeson, E.A., Cann, J.P., and Morris, G.J. (1984). Maintenance of algae and protozoa. In Kirsop, B.E., and Snell, J.J.S. (Eds.), *Maintenance of Microorganisms*. Academic Press, London, pp. 131–160.

Lehman, J.T. (1976). Photosynthetic capacity and luxury uptake of carbon during phosphate limitation in *Pediastrum duplex* (Chlorophyceae). *J. Phycol.*, 12: 190–193.

Lewin, R.A. (1959). The isolation of algae. *Rev. Algol.* (new series), 3: 181–197.

Lokhorst G.M. (2003). The genus *Tribonema* (Xanthophyceae) in the Netherlands. An integrated field and culture study. *Nova Hedwigia,* 77: 19–53.

McDaniel, H.R., Middlebrook, J.B., and Bowman, R.O. (1962). Isolation of pure cultures of algae from contaminated cultures. *Appl. Microbiol.*, 10: 223.

Melkonian, M. (1990). Phylum Chlorophyta: Class Chlorophyceae. In *Handbook of Protoctista*, Margulis, L., Corliss, J.O., Melkonian, M. and Chapman, D.J., (Eds.), 600–607. Boston: Jones and Bartlett Publishers.

Morel, F.M.M., Westall, J.C., Reuter, J.G., and Chaplick, J.P. (1975). Description of the algal growth media "Aquil" and "Fraquil." Water Quality Laboratory, Ralph Parsons Laboratory for Water Resources and Hydrodynamics, Cambridge, Massachusetts: Massachusetts Institute of Technology, Technical Report 16, p. 33.

Moreno-Garrido, I. (2008). Microalgae immobilization: Current techniques and uses. *Bioresour. Technol.*, 99: 3949–3964.

Mutanda, T., Ramesh, D., Karthikeyan, S., Kumari, S., Anandraj, A., and Bux, F. (2011) Bioprospecting for hyper-lipid producing microalgal strains for sustainable biofuel production. *Bioresource Technol.*, 101(1): 57–70.

Nalewajko, C., Lee, K., and Olaveson, M. (1995). Responses of freshwater algae to inhibitory vanadium concentrations: The role of phosphorus. *J. Phycol.*, 31: 332–343.

Nichols, H. W. (1973). Growth media—freshwater. In Stein, J. (Ed.) *Handbook of Phycological Methods, Culture Methods and Growth Measurements*, Cambridge, UK: Cambridge University Press, pp. 7–24.

Noël, M.-H., Kawachi, M., and Inouye, I. (2004). Induced dimorphic life cycle of a coccolithophorid, *Calyptrosphaera sphaeroidea* (Prymnesiophyceae, Haptophyta). *J. Phycol.*, 40: 112–129.

Norton, T.A., Melkonian, M., and Andersen, R.A. (1996). Algal biodiversity. *Phycologia*, 35(4): 308–326.

Olaizola, M. (2003). Commercial development of microalgal biotechnology: From the test tube to the market place. *Biomol. Eng.*, 20: 459–466.

Petersson, F., Nilsson, A., Holm, C., Jönsson, H., and Laurell, T. (2004). Separation of lipids from blood utilizing ultrasonic standing waves in microfluidic channels. *Analyst,* 129: 938–943.

Raja, R., Hemaiswarya, S., Kumar, N.A., Sridhar, S., and Rengasamy, R. (2008). A perspective on the biotechnological potential of microalgae. *Crit. Rev. Microbiol.*, 34(2): 77–88.

Reardon, E.M., Price, C.A., and Guillard, R.R.L. (1979). Harvest of marine microalgae by centrifugation in density gradients of Percoll. In Reid, E. (Ed.), *Cell Populations. Methodological Surveys (B) Biochemistry*. Vol. 8. John Wiley & Sons, New York, pp. 171–175.

Reckermann, M. (2000). Flow sorting in aquatic ecology. *Science,* 64: 235–246.

Richmond, A. (1983). *Handbook of Microalgal Mass Culture*, Boca Raton, Florida: CRC Press.

Richmond, A., (2004). *Handbook of Microalgal Culture: Biotechnology and Applied Phycology*, Blackwell Science Ltd., Malden, MA, p. 566.

Rippka, R. (1988). Isolation and purification of cyanobacteria. *Method Enzymol.*, 167: 3–27.

Rippka, R., De Reuelles, J., Waterbury, J.B., Herdman, M., and Stainer, R.Y. (1979). Generic assignments, strains histories and properties of pure cultures of cyanobacteria. *J. Gen. Microbiol.,* 111: 1–161.

Rogerson, A., De Freitas, A.S.W., and McInnes, A.C. (1986). Observations on wall morphogenesis in *Coscinodiscus asteromphalus* (Bacillariophyceae). *Trans. Am. Microsci. Soc.,* 105: 59–67.

Round, F.E. (1984). *The Ecology of Algae.* Cambridge University Press, Cambridge, UK, p. 664.

Santarius, K.A. (1996). Freezing of isolated thylakoid membranes in complex media. X. Interactions among various low molecular weight cryoprotectants. *Cryobiology,* 33: 118–126.

Sensen, C., Heimann, K., and Melkonian, M. (1993). The production of clonal and axenic cultures of microalgae using fluorescence activated cell sorting. *Eur. J. Phycol.,* 28: 93–97.

Sheehan, J., Dunahay, T., Benemann, J., and Roessler, P. (1998). A look back at the U.S. Department of Energy's aquatic species program – Biodiesel from algae. Available <http://www.nrel.gov/docs/fy04osti/34796.pdf>.

Shevkoplyas, S.S., Yoshida, T., Munn, L.L., and Bitensky, M.W. (2005). Biomimetic autoseparation of leukocytes from whole blood in a microfluidic device. *Anal. Chem.,* 77: 933–937.

Sieracki, M.E., Poulton, N.J., and Crosbie, N. (2005). Automated isolation techniques for microalgae. In R.A. Anderson (Ed.), *Algal Culturing Techniques,* Elsevier Academic Press, Burlington, MA, pp. 101–116.

Sinigalliano, C.D., Winshell, J., Guerrero, M.A., Scorzetti, G., Fell, J.W., Eaton, R.W., Brand, L., and Rein, K.S. (2009). Viable cell sorting of dinoflagellates by multiparametric flow cytometry. *Phycologia,* 48: 249–257.

Spolaore, P., Joannis-Cassan, C., Duran, E., and Isambert, A. (2006). Commercial applications of microalgae. *J. Biosci. Bioeng.,* 101: 87–96.

Steup, M., and Melknonian, M. (1981). C-1,4-Glucan phosphorylase forms in the green alga *Eremosphaera viridis. Physiol. Plant.,* 51: 343–348.

Takahashi, K., Hattori, A., Suzuki, I., Ichiki, T., and Yasuda, K. (2004). Non-destructive onchip cell sorting system with real-time microscopic image processing. *J. Nanobiotechnol.,* 2: 5.

Tang, Y.Z., and Dobbs, F.C. (2007). Green autofluorescence in dinoflagellates, diatoms, and other microalgae and its implications for vital staining and morphological studies. *Appl. Environ. Microbiol.,* 73: 2306–2313.

Taylor, R., and Fletcher, R.L. (1998). Cryopreservation of eukaryotic algae—A review of methodologies. *J. Appl. Phycol.,* 10: 481–501.

The World Conservation Union. (2010). IUCN Red List of Threatened Species. Summary Statistics for Globally Threatened Species. Table1: Numbers of threatened species by major groups of organisms (1996–2010). <http://www.iucnredlist.org/documents/summarystatistics/2010_1RL_Stats_Table 1.pdf>.

Throndsen, J. (1978). Preservation and storage. In Sournia, A.A., (Ed.), *Phytoplankton Manual: Monographs on Oceanographic Methodology 6.* UNESCO, Paris, pp. 69–74.

Vonshak, A. (1986). Laboratory techniques for the cultivation of microalgae. In Richmond, A. (Ed.). CRC *Handbook of Microalgal Mass Culture.* Boca Raton, Florida: CRC Press. pp. 117–145.

Walne, P. R. (1970). Studies on food value of nineteen genera of algae to juvenile bivalves of the genera *Ostrea, Crassostrea, Mercenaria and Mytilus. Fish. Invest. Lond. Ser. 2.,* 26(5): 1–62.

Warren, A., Day, J.G., and Brown, S. (2002). Cultivation of protozoa and algae. In Hurst, C.J., Crawford, R.L., Knudsen, G.R., McInerney, M.J., and Stezenbach, L.D. (Eds.), *Manual of Environmental Microbiology, 2nd ed.* ASM Press, Washington, D.C., pp. 71–83.

Watanabe, M.M., Kawachi, M., Hiroki, M., and Kasai, F. (2000). *NIES Collection List of Strains. Sixth Edition, 2000, Microalgae and Protozoa.* Tsukuba, Japan: Microbial Culture Collections, National Institute for Environmental Studies, p. 159.

Waterbury, J.B., Watson, S.W., Valois, F.W., and Franks, D.G. (1986). Biological and ecological characterization of the marine unicellular cyanobacterium *Synechococcus. Can. Bull. Fish. Aquatic Sci.*, 214: 71–120.

Wetzel, R.G., and Likens, G.E. (2000). *Limnological Analyses, third edition.* Springer-Verlag, New York, p. 429.

Wiedeman, V.E., Walne, P.L. and Trainor, F.R. (1964). A new technique for obtaining axenic cultures of algae. *Can. J. Bot.*, X: 958–959.

Wilkie, A.C., and Mulbry, W.W. (2002). Recovery of dairy manure nutrients by benthic freshwater algae, *Bioresour. Technol.*, 84(1): 81–91.

Wilkie, A.C., Edmundson, S.J., and Duncan, J.G. (2011). Indigenous algae for local bioresource production: Phycoprospecting. *Energy for Sustainable Develop.*, 15: 365–371.

World Agricultural Supply and Demand Estimates. (2012). WASDE-503, February 9, 2012. U.S. Department of Agriculture, World Agricultural Outlook Board, Washington, D.C.: 2012, <http://www.usda.gov/oce/commodity/wasde/latest.pdf>.

Wu, H., Volponi, J.V., Oliver, A.E., Parikh, A.N., Simmons, B.A., and Singh, S. (2011). In vivo lipidomics using single-cell Raman spectroscopy. *Proc. Natl. Acad. Sci. USA,* 108(9): 3809–3814.

Yamada, M., Kasim, V., Nakashima, M., Edahiro, J., and Seki, M. (2002). Continuous cell partitioning using an aqueous two-phase flow system in microfluidic devices. *Biotechnol. Bioeng.*, 78(4): 467–472. Retrieved from <http://www.ncbi.nlm.nih.gov/pubmed/15459911>.

Yang, H.L., Lu, C.K., Chen, S.F., Chen, Y.M., and Chen, Y.M. (2010). Isolation and characterization of Taiwanese heterotrophic microalgae: Screening of strains for docosahexaenoic acid (DHA) production. *Mar. Biotechnol.*, 12(2): 173–185.

Zhou, W., Li, Y., Min, M., Hu, B., Chen, P., and Ruan, R. (2011). Local bioprospecting for high-lipid producing microalgal strains to be grown on concentrated municipal wastewater for biofuel production. *Bioresource Technology,* 102(13): 6909–6919.

4 Enumeration of Microalgal Cells

Taurai Mutanda and Faizal Bux
Institute for Water and Wastewater Technology
Durban University of Technology
Durban, South Africa

CONTENTS

4.1 INTRODUCTION

The estimation of a microalgal population size is no easy task due to the microscopic size of the cells. Consequently, it is impossible to physically count them with the naked eye. The size of microalgae falls within the size of other microbes (e.g., bacteria) and, as a result, most of the methods used for microbial cell counting are also applicable to microalgae. In general, conventional microbiological protocols are available and are sufficient for cell enumeration despite the proliferation of modern and advanced techniques. It is recommended that the researcher choose a method after assessing the costs involved because some of the latest methods require very sophisticated and expensive equipment.

According to Caron et al. (2003), "the identification and enumeration of microorganismal species in natural aquatic assemblages is an essential prerequisite for ecological studies of these populations." Effective ecological studies of populations of colonial freshwater phytoplankton species are hampered by a lack of methods for cell enumeration (Box, 1981). Closely related microalgal groups must be accurately distinguished, and this is very crucial when these species pose health and environmental risks (Caron et al., 2003).

Microalgal cell population size is important when studying growth kinetics. It is also important to know the amounts of biochemical constituents such as pigments (e.g., chlorophyll-a). Microalgal cells can be enumerated directly using techniques such as light absorption and/or indirectly via surrogate determination of dry or wet biomass and/or measurement of cell components such as organic nitrogen, phosphorus, etc. (Guillard and Sieracki, 2005).

However, despite their common usage, direct methods and gravimetric measurements of microalgal cell dry weights are tedious, time consuming, and prone to errors that may exceed ±10% (Elnabarawy and Welter, 1984). Techniques for the estimation of the abundance of microalgal cells in natural microalgal samples are improving at a fast pace with the advent of electronic counting methods. Despite the progress made in the development of these advanced and sophisticated techniques, they still have major drawbacks, such as cost and the requirement of highly skilled personnel to operate the equipment. This chapter describes some of the popular methods that are available for the enumeration of microalgal cultures. The methods that are widely used are spectrophotometry, dry weight determinations, light microscopy, haemacytometry, and flow cytometry.

4.2   SAMPLING AND CULTURING MICROALGAE

It is immensely difficult to sample and accurately count microalgal cells that grow attached to the substratum or culture vessel. The cells are scraped from the surface using a spatula to dislodge them into a suspension. The cells are then carefully homogenized so as to separate the cell clumps, and this can result in the death of some cells. The cells that are clumped are indistinguishable from single cells, and therefore cell enumeration may be inaccurate (Guillard and Sieracki, 2005). Colonies of *Microcystis aeruginosa* were separated to a homogenous cell suspension by alkaline hydrolysis with 0.01 M KOH at 80°C for about 30 minutes, but higher molarities of KOH resulted in cell loss (Box, 1981). In addition, complete separation of colonies to single cells was achieved by heating at 80°C for 30 minutes followed by 30 seconds of vortex-mixing (Box, 1981). Furthermore, in these studies, sonication (20 KHz, *c.* 50W) did not completely reduce the colonies to single cells, and this procedure resulted in cell death on some occasions (Box, 1981).

It has been reported that the efficiency of the method for separating the colonies into a single-cell suspension relies on the microalgal cells used (Box, 1981). The accuracy of enumerating microalgal cells displaying colonial growth is hindered if there is no uniformity in the number of cells per colony. In addition, the cells cannot be sufficiently distinguished from each other, thus compromising accuracy. The enumeration of dividing microalgal cells is subject to inherent inaccuracies.

4.3   MICROALGAL PRESERVATION

Microalgal cells must be preserved after sampling, most preferably in Lugol's iodine solution before counting, to arrest the movement of some live cells around the counting chambers, for example, flagellates and diatoms, and therefore hampering accurate enumeration. Other methods of preservation, such as the use of aldehydes, saline ethanol, and freezing, can also be adopted for a wide range of phytoplankton. Microalgal preservation is recommended so as to prevent cell disintegration and to avoid any changes in cell population size due to zooplankton grazing (Hotzel and Croome, 1999).

4.4 ENUMERATION METHODS

There are several methods available for the enumeration of microalgal cells. However, due to the small size of the microalgal cells, most methods are not very accurate. In addition, some of the methods described here require sophisticated and expensive equipment, which is beyond the reach of some research laboratories. The choice of counting device depends on culture density, the size and shape of the cells or colonies being counted, and the presence and amount of extracellular threads, sheaths, or dissolved mucilage, which can influence the filling of the counting chamber (Guillard and Sieracki, 2005). Moreover, the method to be adopted for a particular sample will therefore depend on other factors such as detection range, costs, sample throughput, health and safety considerations, *inter alia.*

4.4.1 SPECTROPHOTOMETRIC ANALYSIS

The use of a spectrophotometer for indirectly measuring microalgal cell density is done in conjunction with other methods, for example, gravimetric and counting chambers for the calibration curve (see below). The wavelength for the determination of microalgal biomass is in the visible range of the electromagnetic spectrum (400 to 700 nm). It is recommended to do a calibration curve with samples of known microalgal cell numbers so as to extrapolate cell numbers from the standard curve. Spectrophotometric analysis is an easy quantification method although not very accurate because some other non-microalgal particles such as artifacts and dissolved and suspended solids may contribute to light absorption. However, it is recommended to ensure that the samples to be analyzed are free of other contaminants prior to analysis because this method does not discriminate noncellular materials from microalgal cells. One major drawback of spectrophotometric analysis is that culture conditions greatly affect chlorophyll content, which in turn determines absorbance. Microalgal cells grown in dark conditions have higher chlorophyll contents as compared to cells grown under high light intensity.

The use of light absorption is suitable for the estimation of microalgal population size rather than the determination of the actual number of individual cells.

4.4.2 GRAVIMETRIC ANALYSIS

The quantitative determination of biomass by gravimetric analysis is easy, cost effective, and the equipment required for this purpose is not very expensive. However, this technique determines the weight of the whole biomass and does not discriminate individual cells in the media. In general, wet or dry weights of the biomass can be measured, although wet weight determination is not very accurate. The microalgal cells are harvested by centrifugation and the pellet washed twice with isotonic ammonium formate (NH_4HCO_2, 2%) to remove suspended residual salts (Valenzuela-Espinoza et al., 2002). The advantage of using ammonium formate is that it does not leave any residue as it decomposes to volatile compounds during the drying process. An empty watch glass is weighed and the microalgal sample pellet transferred to the watch glass. The biomass is dried in an oven at 60°C for 12 hours. After drying, the weight of the watch glass plus the dry biomass is determined, and the net dry cell weight (DCW) is calculated using Equation (4.1):

$$\text{DCW (mg L}^{-1}) = [\{\text{Watch glass (mg)} + \text{Dry biomass (mg)}\} \\ - \text{Watch glass (mg)}]/\text{Volume (L)} \quad (4.1)$$

4.4.3 COUNTING CHAMBERS

The counting chamber methods are well established and frequently applied for microalgal enumeration due to their low cost and easy application. The three common types of counting chamber methods for microalgae enumeration are the (1) Sedgewick–Rafter counting slide, (2) Palmer–Maloney counting slide, and (3) haemocytometer counting slide (LeGresley and McDermott, 2010). The three methods require samples with high cell densities. The presence of contaminating particles in the same size range as the algae and failure of cells to separate after cell division may be possible sources of erroneous counts (Coutteau, 1996). Table 4.1 compares the merits and drawbacks as well as fundamentals of the three counting chamber methods.

4.4.4 FLOW CYTOMETRY

Flow cytometry (FCM) is an automated cell counting technique that captures the fluorescence and scatter properties of the microalgal cells. The major advantage of automated cell counting techniques over optical microscopy is that they minimize the errors associated with human counting (Marie et al., 2005). The use of this highly sophisticated technique is hindered by the cost of the equipment as well as the requirement of highly skilled and trained personnel to operate the instrument. In addition, this technique is also plagued by limited sensitivity at lower microalgal cell concentrations. The solid-phase cytometer method for conducting total direct counts of bacteria is less biased and has performed significantly better than any of the microscopic methods (Lisle et al., 2004). Basic image analysis methods do not generally discriminate between phytoplankton and other material such as detritus and sediment in samples, thereby presenting a problem in the application to routine field samples. This technique may be more useful for the analysis of cultures and mono-specific high-density blooms (Karlson et al., 2010).

4.5 CONCLUSION

Microalgal enumeration can be tedious and cumbersome due to the small size of microalgal cells. Furthermore, this is exacerbated by the prohibitive cost of the available sophisticated equipment. To date, however, microalgal enumeration has been accomplished by gravimetric analysis, counting chambers, and flow cytometry.

TABLE 4.1
Comparison of the Fundamentals of the Counting Chamber Methods

Counting Chamber	Sedgewick–Rafter	Palmer–Maloney	Haemocytometer
Scope	Cultures and high cell numbers	Cultures and high cell numbers as in bloom situations	Cultures and extremely high cell concentrations of small organisms
Detection range	1000 cells L⁻¹ Limit of Detection (LOD)	10,000 cells L⁻¹ (LOD)	10,000,000 cells L⁻¹ (LOD)
Advantages	A rapid estimate of cell concentrations	A rapid estimate of high cell concentrations	A rapid estimate of extremely high cell concentrations
Drawbacks	Accurate results only when sample contains high algal cell densities	Accurate results only when sample contains very high algal cell densities	Accurate results only when sample contains extremely high algal cell densities
Type of training needed	Method-easy to learn and use; highly trained taxonomist needed for verification of species identification	Method-easy to learn and use; highly trained taxonomist needed for verification of species identification	Method-easy to learn and use; highly trained taxonomist needed for verification of species identification
Essential equipment	Compound Microscope Cover slips Pipettes Sedgewick–Rafter slides	Compound Microscope Cover slips Pipettes Palmer–Maloney slides	Compound Microscope Pipettes Haemocytometer slide with cover glass.
Equipment cost	Compound microscope: £2500/US$3250 Sedgewick–Rafter slides: Perspex: £50/US$65 Glass: £166/US$213	Compound microscope: £2500/US$3250 Palmer–Maloney slides: Ceramic-£60/US$80 Stainless steel-£170/US$230	Compound microscope: £2500/US$3250 Haemocytometer slide: £200/US$230
Consumables, cost per sample	£1/US$1.3	£1/US$1.3	£1/US$1.3
Processing time/sample	20 min	5 min	5 min
Analysis time/sample	This depends on the sample density	10–30 min/sample depending on cell density	<20 min/sample depending on the sample density
Sample/throughput/person/day	This depends on the sample density	14–30 min/sample depending on the sample density	<30 min dependent on target species
Samples processed in parallel	Only one sample at a time	Only one sample at a time	Only one sample at a time
Health and safety issues	Dependent on preservative used	Dependent on preservative used	Dependent on preservative used

Source: Adapted from LeGresley and McDermott, 2010.

The invention and development of advanced techniques such as fluid imaging alleviates the drawbacks of flow cytometry.

ACKNOWLEDGMENTS

The authors hereby acknowledge the National Research Foundation (South Africa) for financial contribution.

REFERENCES

Box, J.D. (1981). Enumeration of cell concentrations in suspensions of colonial freshwater microalgae, with particular reference to *Microcystis aeruginosa*. *British Phycological Journal*, 16: 153–164.

Caron, D.A., Dennett, M.R., Moran, D.M., Schaffner, R.A., Lonsdale, D.J., Gobler, C.J., Nuzzi, R., and McLean, T.I. (2003). Development and application of a monoclonal-antibody technique for counting *Aureococcus anophagefferens*, an alga causing recurrent brown tides in the Mid-Atlantic United States. *Applied and Environmental Microbiology*, 69(9): 5492–5502.

Coutteau, P. (1996). Microalgae. In *Manual on the Production and Use of Live Food for Aquaculture* (Eds., P. Lavens and P. Sorgeloos), FAO Fisheries Technical Paper Series 36, Belgium, pp. 7–48.

Elnabarawy, M.T., and Welter, A.N. (1984). A technique for the enumeration of unicellular algae. *Bulletin of Environmental Contamination and Toxicology*, 32: 333–338.

Guillard, R.R.L., and Sieracki, M.S. (2005). Counting cells in cultures with the light microscope. In *Algal Culturing* Techniques (Ed., R.A. Andersen), Elsevier Academic, Burlington, MA, pp. 239–252.

Hotzel, G. and Croome, R. (1999). *A Phytoplankton Methods Manual for Australian Freshwaters.* LWRRDC Occasional Paper 22/99.

Karlson, B., Godhe, A., Cusack, C., and Bresnan, E. (2010). Introduction to methods for quantitative phytoplankton analysis. In *Microscopic and Molecular Methods for Quantitative Phytoplankton Analysis* (Eds., Karlson, B., Cusack, C., and Bresnan, E.). Intergovernmental Oceanographic Commission, UNESCO, Paris, pp. 5–12.

LeGresley, M., and McDermott, G (2010). Counting chamber methods for quantitative phytoplankton analysis – Haemocytometer, Palmer–Maloney cell and Sedgewick–Rafter cell. In *Microscopic and Molecular Methods for Quantitative Phytoplankton Analysis* (Eds., Karlson, B., Cusack, C., and Bresnan, E.). Intergovernmental Oceanographic Commission, UNESCO, Paris, pp. 25–30.

Lisle, J.T., Hamilton, M.A., Willse, A.R., and McFeters, G.A. (2004). Comparison of fluorescence microscopy and solid-phase cytometry methods for counting bacteria in water. *Applied and Environmental Microbiology,* 70(9): 5343–5348.

Marie, D., Simon, N., and Vaulot, D. (2005). Phytoplankton cell counting by flow cytometry. In R.A. Andersen (Ed.), *Algal Culturing Techniques*, pp. 253–268. Elsevier Academic, Burlington, MA.

Valenzuela-Espinoza, E., Millan-Nunez, R., and Nunez-Cebrero, F. (2002). Protein, carbohydrate, lipid and chlorophyll *a* content in *Isochrysis* aff. *galbana* (clone TIso) cultured with a low cost alternative to the f/2 medium. *Aquacultural Engineering*, 25(4): 207–216.

5 Microalgal Cultivation Reactor Systems

Melinda J. Griffiths
Centre for Bioprocess Engineering Research
University of Cape Town, South Africa

CONTENTS

5.1 INTRODUCTION

One of the most important factors in achieving economically and environmentally feasible commercial-scale production of microalgae is the development of cost-effective, sustainable culture systems (Borowitzka, 1999; Richmond, 2000). The design of the cultivation system influences the environmental conditions experienced by the cells, which in turn determine the productivity (Greenwell et al., 2010).

Improving productivity is key to achieving economic viability in large-scale, outdoor cultures (Lee, 2001).

Microalgal cultivation has been carried out in a variety of vessels, ranging from natural open lakes and ponds to highly complex and controlled photobioreactors (PBRs). Typically, the term *photobioreactor* has been used to refer to closed systems exclusively; however, by definition, open systems are also PBRs. A bioreactor is a container in which living organisms are cultivated and carry out biological conversions (e.g., biomass or product formation) (Grobbelaar, 2009). A PBR is a reactor in which organisms that obtain energy from light, such as algae, plants, and certain microbial cells (phototrophs), are used to carry out reactions (Mata et al., 2010). Each type has advantages and disadvantages, but the overall goals are similar. In the design of commercial algae cultivation systems, the aim is to achieve:

- Optimal volumetric and/or areal productivity
- Efficient conversion of light energy to product
- Consistency and reliability of production
- Cost effectiveness

Effective reactor design requires knowledge of both algal physiology and reactor engineering, such as aspects of hydrodynamics and mass transfer (Ugwu et al., 2008). Section 5.2 outlines the key requirements for algal growth and how these relate to design considerations of the cultivation system. Section 5.3 describes the range of open and closed systems that have been used for microalgal cultivation. These are compared with respect to a range of attributes in Section 5.4.

5.2   GROWTH REQUIREMENTS AND DESIGN PARAMETERS

For optimal microalgal growth, several environmental parameters (e.g., temperature, light intensity, pH, and nutrient concentrations) must be kept within narrow physiological limits. The reactor system is critical in the provision and maintenance of a favorable growth environment (Pulz, 2001). Hence, reactor design requires knowledge of aspects of algal physiology, such as the morphology, nutrient requirements, and stress tolerance of the species to be grown. Some of the requirements for microalgal growth are listed in Table 5.1, along with the consequences of under- or overprovision, and the relevant reactor design features.

5.2.1   LIGHT

Light is the principal limiting factor in the culture of photosynthetic organisms (Pulz, 2001); therefore, the intensity and utilization efficiency of the light supply are critical in reactor design (Kumar et al., 2010). The photosynthetic activity of microalgae changes in response to light intensity in three distinct regions. At low light intensities, cells are light limited and the photosynthetic rate increases with increasing irradiance. Once cells become light saturated, the rate of photon absorption exceeds the rate of electron turnover in Photosystem II (PS II), and there is no further increase in the photosynthetic rate with increasing light intensity. Once irradiance increases above

TABLE 5.1

Key Requirements for Algal Growth in Relation to PBR Design

Key Requirement	Consequences if Too Low	Consequences if Too High	Function of
Light	Insufficient for photosynthesis, slow growth	Photo-inhibition, photo- and oxidative damage	Reactor surface:volume ratio Geometry, orientation, and inclination of reactor Material and thickness of reactor walls Culture depth and density Mixing
Temperature	Slow growth, dormancy	Cell death	Heat input (ambient temperature, solar radiation, angle to sun, shading, heat generation by algal metabolism) Heat dissipation (evaporation, airflow, heating/cooling mechanisms)
Nutrient provision	Growth inhibition	Toxicity	Media composition CO_2 provision and O_2 removal (mass transfer, sparging and degassing mechanisms, gas concentration and flow rate, headspace, gas holdup volume) Mixing
Mixing	Poor mass transfer Biomass settling Anaerobic zones	Shear stress High energy use	Reactor geometry, mixing technique (e.g., mechanical, air flow, gravity flow)

a certain point, cells become photo-inhibited due to damage to the photosynthetic apparatus, and the photosynthetic rate declines with further increases in irradiance (Chisti, 2007; Grobbelaar, 2009). In most algae, photosynthesis is saturated at about 1,700 to 2,000 $\mu mol\ m^{-2}s^{-1}$, while some plankton are photo-inhibited at much lower levels (130 $\mu mol\ m^{-2}s^{-1}$). Photo-inhibition occurs rapidly; irreversible destruction can occur in a few minutes, exceeding 50% damage after 10 to 20 minutes (Pulz, 2001).

In dense algal cultures under high irradiance (e.g., mid-day sunlight), it is likely that the illumination at the culture surface will be sufficient to induce photo-inhibition, while that a few centimeters below the surface will be insufficient for growth. The light conditions experienced by an individual cell within a reactor are constantly changing as a function of

- Culture depth or optical cross-section (the deeper or wider the culture vessel, the longer cells spend in low light conditions)

- Biomass concentration or areal density
- Turbulence induced by mixing (influences light-dark cycling as cells move in and out of the photic volume) (Grobbelaar, 2009).

The requirement for optimal light provision to all cells places unique constraints on the geometry of reactors. As light enters a culture surface, it is absorbed and scattered by the cells, particulate matter, colored or chemical substances, as well as the water itself (Grobbelaar, 2009). As cells at the surface absorb light, they shade those below them. Due to this mutual shading effect, light intensity decreases with culture depth. Light does not penetrate more than a few centimeters into a dense algal culture; therefore, optical depth must be minimized. Reactor scale-up is based on reactor surface area rather than volume, as in the case of heterotrophic fermentations, and the surface-area-to-volume ratio is a critical parameter (Scott et al., 2010). Reactor design is a trade-off between maintaining a shallow depth, or thin optical cross-section, and the increased cost of reactor materials, decreased efficiency of mixing, and greater land area involved.

The density of the culture determines the attenuation of light with distance from the reactor surface. Given a certain reactor path length and light intensity, there will be a corresponding optimal cell density. Below the optimal areal density, all cells are exposed to excess light, and above optimal density, a significant proportion of the culture is in the dark (Grobbelaar, 2009). At the optimal density, given sufficient mixing, all cells are subject to equal light-dark fluctuations. Maximum photosynthetic efficiency occurs in relatively dilute cultures. The increase in productivity achieved by maintaining the optimal cell density for light provision must be balanced against the costs of the increased reactor volume and harvesting capacity required to process large volumes of dilute cell suspensions. In addition, a high volumetric yield does not necessarily mean that incident light is being most efficiently used. This is measured by areal yield, not volumetric yield, and for this there is an optimal areal cell density as well as cell concentration (Richmond, 2000).

Microalgal cells can become acclimated to high or low light conditions. In an effort to balance the activity of the light and dark photosynthetic reactions, cells modulate their light-harvesting capacity (e.g., through adjusting the number of PS II reaction centers and the pigment concentration), depending on the ambient light intensity. The process of photo-adaptation takes 10 to 40 minutes (Pulz, 2001). Due to the fact that a culture may become acclimated to prevailing light conditions, the optimal biomass concentration is different at high and low irradiance. It is therefore impossible to operate at a single optimum cell concentration when a range of irradiance occurs over the course of the day (Lee, 2001).

It has been postulated that there is a phenomenon known as the flashing light effect that leads to increased productivity at certain frequencies of light-dark cycling. Exposing cells to very short cyclic periods of light and darkness could counterbalance the two extremes of light over-saturation and inhibition. However, the effect of flashing light is very difficult to separate experimentally from the effects of the increased turbulence required to generate faster light-dark cycling (Grobbelaar, 2009). It is clear that enhanced mixing, up to a point at which cell damage begins to occur, is beneficial to optimal light provision by creating an average light intensity across the reactor volume by rotating cells between the light and dark phases of the reactor.

In addition to individual reactor design, the configuration of multiple reactor units can be designed for optimal light distribution. For example, placing plate reactors very close together dilutes strong light, which leads to an increase in photosynthetic efficiency. Overlapping tubular systems can also be used to dilute strong sunlight. However, the benefits of high photosynthetic efficiency may be offset by the increased cost of reactor hardware (Richmond, 2000).

The use of internal illumination can remove some of the surface-area-to-volume constraint on bioreactor design (Ugwu et al., 2008). Both natural and artificial light sources could be utilized, either using optic fibers to distribute solar energy inside the PBR or placing waterproof artificial illumination internally. Artificial lighting has the advantage that it can be used to supplement the light supply at night or during cloudy days. However, it adds to the operational cost and energy input; therefore, higher biomass yields are crucial (Ugwu et al., 2008; Kumar et al., 2010).

5.2.2 Temperature

Along with light intensity, temperature is one of the most difficult parameters to optimize in large-scale outdoor culture systems. Fluctuations in temperature, both daily and seasonally, can lead to significant decreases in productivity. The optimal growth temperature for microalgae is species specific, but often in the region of 20°C to 30°C (Chisti, 2008). Many algal species can tolerate temperatures of up to 15°C lower than their optimum, with reduced growth rates, but a temperature of only a few degrees higher than optimal can lead to cell death (Mata et al., 2010). The net efficiency of photosynthesis declines at high temperature as the rate of respiration rises significantly, while the increased flux through the Calvin cycle is moderate. This effect is worsened by the fact that CO_2 becomes less soluble at elevated temperatures, more rapidly than O_2 (Pulz, 2001).

Low seasonal, morning, and evening temperatures can lead to significant losses in productivity, although low nighttime temperatures are potentially advantageous due to a reduction in the respiration rate. As much as 25% of the biomass produced during daylight hours can be lost at night due to respiration (Chisti, 2007). Cool nighttime temperatures can minimize this loss.

Closed reactor systems almost always require some form of temperature control. They often suffer from overheating during hot days when temperatures inside the reactor can reach in excess of 50°C. Heat exchangers or evaporative water-cooling systems may be employed to counteract this (Mata et al., 2010). The culture system can also be placed inside a greenhouse, or contacted with water to minimize temperature fluctuations (Chisti, 2007). Closed PBRs are sometimes floated, either whole or just the solar collector, in a temperature-modulating water bath. Double-walled reactors with part of the liquid volume used for heating and cooling have been devised (Ugwu et al., 2008), although all such modifications add to the cost of production.

There is a relationship between temperature and light availability. Exposure to a rapid increase in light intensity when the temperature is below optimum (as occurs in the early morning in outdoor cultures) can lead to photo-inhibitory stress as cells are too cold to process incoming photons, thereby reducing photosynthetic efficiency for a good part of the morning (Vonshak, 1997). Low temperatures are therefore particularly suboptimal in the early morning, and any efforts to employ heat reactors should be concentrated just before dawn.

5.2.3 NUTRIENT PROVISION

Optimal supply of nutrients, mainly carbon, nitrogen, and phosphorous, along with various other macro- and micronutrients required for algal growth, is a prerequisite for high growth rates. Deficiencies in any nutrient cause disturbances in metabolism, physiological changes, and decreased productivity (Pulz, 2001). The supply of nutrients to the culture is relatively simple, but the supply of nutrients to individual cells depends on efficient mass transfer, which is related to mixing and gas sparging (Grobbelaar, 2009). Nutrients are also a significant cost in microalgal cultivation; therefore, design of the reactor system to allow for efficient recycling of culture medium is essential (Greenwell et al., 2010).

Nutrients, with the exception of light and carbon, are generally provided in the liquid growth medium. Carbon is a major constituent of algal cells (often comprising 50% of the dry weight), usually obtained from carbon dioxide (CO_2) gas (Chisti, 2007). The concentration of CO_2 in air (0.04%) is suboptimal for plant growth; therefore, for optimal productivity, CO_2-enriched air must be supplied (Pulz, 2001). CO_2 may be available from flue gas or other waste gas streams, but the cost of gas compression and extensive sparging systems for arrays of PBRs is significant. The location of large algal plants sufficiently near the source, along with the safety concerns of large-scale distribution of flue gas (low in O_2 and high in CO_2, NO_X, and SO_X) at ground level, could present its own challenges (Scott et al., 2010).

In cases of carbon limitation, the efficiency of mass transfer of CO_2 from gas to liquid form in the culture medium becomes critical to productivity. Certain algal species can grow heterotrophically or mixotrophically, in which case all or some of the carbon and energy requirements can be supplied from an organic carbon source such as glucose or acetate (Lee, 2001). The use of heterotrophy can reduce the dependence of productivity on light and CO_2 supply, which releases some of the key constraints on reactor design (Pulz, 2001). Heterotrophic cultivation of microalgae in sterilizable fermenters has achieved some commercial success, although biomass productivity has yet to match that of yeast and other heterotrophic organisms (Lee, 2001).

Mixing and nutrient concentrations are also linked to pH control. Mixing promotes reactions of CO_2 with H^+, OH^-, H_2O, and NH_3 in the medium, which affect the pH and hence CO_2 uptake rates (Kumar et al., 2010). The pH increases along the length of tubular reactors due to consumption of CO_2. In long reactors, CO_2 injection points may be necessary to prevent a rise in pH above optimal levels (Chisti, 2007). CO_2 addition is commonly controlled by feedback from a pH meter (Carvalho et al., 2006).

The removal of toxic metabolites is also critical to the efficiency of growth and photosynthesis. Under high irradiance, oxygen generation in closed PBRs can be up to 10 g $O_2.m^{-3}min^{-1}$. Maximum dissolved oxygen levels should not exceed 400%

saturation (with respect to air-saturated culture) (Chisti, 2007). A build-up of O_2 in the reactor can cause the key carbon-fixing enzyme RuBisCO to bind oxygen instead of carbon dioxide, leading to photorespiration instead of photosynthesis (Dennis et al., 1998). High oxygen concentrations, in addition to intense light, lead to the formation of oxygen radicals that have toxic effects on cells due to membrane damage (Molina Grima et al., 2001; Pulz, 2001). Many algal strains cannot survive in O_2 over-saturated conditions for more than 2 to 3 hours. High temperatures and light intensify the damage (Pulz, 2001). Oxygen build-up limits the maximum length of a closed tubular reactor. Typically, a continuous tube should not exceed 80 m (Molina Grima et al., 2001), although the exact length depends on biomass concentration, light intensity, liquid velocity, and initial O_2 concentration. In a closed reactor, culture must continuously return to a degassing zone, where it is bubbled with air to strip the O_2. The degassing zone is typically optically deep compared with the solar collector, and hence poorly illuminated; thus its volume should be small relative to the solar collector (Chisti, 2007).

In high-density algal cultures, the key challenges in nutrient provision are in mass transfer of CO_2 to cells and O_2 away from cells. Efficient mixing and aeration, without inducing shear stress and requiring excessive energy input, are important parameters. Bubbling of gas through cultures can be used to simultaneously introduce CO_2, strip O_2, and mix the culture broth (e.g., bubble columns and airlift reactors). The overall mass transfer coefficient (kLa) of the reactor is an important parameter in determining the carbon supply. The kLa depends on reactor geometry, agitation rate, sparger type, temperature, mixing time, liquid velocity, gas bubble velocity, and gas holdup (Ugwu et al., 2008).

5.2.4 Mixing

Mixing is of paramount importance in microalgal culture systems as it is directly linked to other key parameters such as light provision, gas transfer, and nutrient provision. Good mixing keeps the cells in suspension, eliminates thermal stratification, determines the light-dark regime by moving cells through an optical gradient, ensures efficient distribution of nutrients, improves gas exchange, reduces mutual shading at the center of the reactor, and decreases photo-inhibition at the surface (Ugwu et al., 2008). Mixing affects the mass transfer rates of dissolved nutrients and gases by reducing the boundary layer between the surface of cells and gas particles and the bulk liquid (Grobbelaar, 2009). The synergistic effect on several parameters at once means that mixing efficiency has a strong effect on growth rate.

One of the major differences between open and closed reactors is the degree of turbulence achieved. Higher turbulences are more easily achieved in closed PBRs with narrow tubes or plates. Mixing in open ponds is typically provided by a paddlewheel or rotating arm. In closed reactors, mixing can be achieved mechanically (by pumping or stirring) or by aeration via a variety of gas transfer systems (e.g., bubble diffusers, pipes, blades, propellers, jet aerators, or aspirators). Stirring is efficient but incurs high mechanical stress. Mixing by gas injection is relatively gentle and efficient, but may require energy intensive gas pressurization. Gas introduced into reactors can serve a number of purposes, including supply of nutrients, control of pH, stripping of O_2, and mixing. Bubbling of CO_2 into the bottom of reactors is generally favored, although it achieves only moderate transfer efficiencies (13% to 20%) due to loss of CO_2 to the atmosphere, fouling of diffusers, and poor mass transfer (Kumar et al., 2010).

Mixing must be almost continuous to prevent settling of biomass (Molina Grima, 1999) and can represent the major energy input into reactor maintenance. High rates of mixing can also impose shear stress on microalgal cells, particularly in filamentous species or those with delicate morphology (Greenwell et al., 2010). Mixing rates are therefore a trade-off among enhanced growth rate, cell damage, and energy requirement.

5.3 CULTIVATION SYSTEMS

A wide variety of open and closed reactor systems have been proposed for microalgal cultivation, possibly reflecting the diversity in the physiology and requirements of different algal species. Ultimately, the overall goal is the continuous maintenance of a desired algal culture under conditions for optimal productivity. High volumetric and areal yields reduce cost by minimizing the reactor volume and land area required, respectively. Important factors in achieving this include (Richmond, 2000):

- Provision of sufficient light, despite daily and seasonal variations and dense algal culture
- Optimal mixing and mass transfer, while avoiding damage to cells by shear stress
- Minimization of deviation from optimal temperature (requires cooling in summer and heating in winter)
- Minimization of dissolved oxygen tension
- Simple cleaning and maintenance
- Minimization of energy input requirements
- Minimization of water use (e.g., evaporation from ponds, evaporative cooling use)
- Low capital and operating costs per unit of harvested product

5.3.1 OPEN SYSTEMS

Historically, the vast majority of commercial production has been carried out in open ponds, and they are still the most widely applied reactor system in industrial microalgal processes (Carvalho et al., 2006). Open systems include natural water bodies, circular ponds, raceway ponds, and cascade systems. The main constraints on growth in open ponds are that it is impossible to control contamination, difficult to keep the culture environment constant, and expensive to harvest the dilute biomass (Carvalho et al., 2006). To maintain a monoculture in open ponds, highly selective culture conditions are necessary in order to guarantee dominance by the desired strain. For this reason, a limited number of species able to tolerate extreme conditions (e.g., *Spirulina,* which grows at high pH, and *Dunaliella,* which requires high salt concentrations) have been successfully grown. Open systems are susceptible to changes in temperature and irradiance due to local weather and climatic conditions, thus making it difficult to maintain optimal growth conditions. Low cell densities are usually obtained, requiring large volumes to be processed during harvesting, and thus increasing the cost of product recovery (Carvalho et al., 2006).

5.3.1.1 Natural Waters

Open systems include natural waters, lakes, and dams where the growth of the microalgae of interest either occurs naturally or is encouraged through addition of nutrients. Harvesting is carried out in situ; for example, *Spirulina* is harvested commercially from Lake Texcoco in Mexico. Although naturally harvested microalgae incur very little cost in cultivation, the productivity and product quality (biologically and toxicologically) cannot be assured (Lee, 2001).

5.3.1.2 Circular Ponds

The first mass culture of microalgae was carried out in circular ponds (Lee, 2001). They are generally simple, round, concrete ponds or dams, mixed by a rotating circular arm fixed in the center of the pond, or by manual stirring. The size of the pond is limited by the strain of the water resistance against the rotating motor. The largest reported pond is 50 m in diameter (Lee, 2001). They are commonly used in Japan, Taiwan, Indonesia, and Ukraine for *Chlorella* cultivation (Lee, 2001; Pulz, 2001).

5.3.1.3 Raceway Ponds

The most commonly used design for commercial microalgal production is the raceway pond. A raceway is an oval-shaped, single- or multiple-loop recirculation channel (Figure 5.1), usually 15 to 20 cm deep, with mixing provided through circulation by a rotating paddlewheel (Pulz, 2001; Brennan and Owende, 2010). Baffles are often placed in the bends of the flow channel to guide the water and facilitate mixing (Chisti, 2007). They are commonly built from concrete or packed earth,

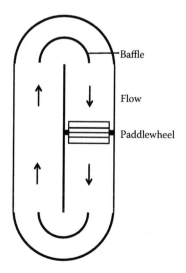

FIGURE 5.1 Schematic diagram of a raceway pond.

and covered with a plastic lining. The largest raceway pond in operation is 5,000 m^2, located at Earthrise Nutritionals, a commercial *Spirulina* producer in Southern California (Spolaore et al., 2006).

By convention, in continuous production, nutrients are introduced in front of the paddlewheel and harvesting takes place behind the wheel (Brennan and Owende, 2010). CO_2 is provided by gas exchange via natural contact with the surrounding air (Singh et al., 2011). Occasionally, submerged aerators are installed to enhance CO_2 absorption. Light provision is by natural sunlight. Ponds can be placed inside covered tunnels to aid in temperature regulation.

Raceway ponds incur relatively low capital investment as well as operational costs. Weekly monitoring is usually sufficient, and the main costs are in the media components and the energy consumed for mixing (Singh et al., 2011). Biomass concentrations are normally in the region of 0.5 g L^{-1}, with a biomass productivity of 10 to 25 g m^{-2}d^{-1} (Sheehan et al., 1998; Lee, 2001). Raceway ponds have been in use since the first commercial microalgal ventures were established, and extensive experience in their design and operation exists. Examples of the productivities obtained with various microalgal species in raceway ponds are given in Table 5.2.

The only open system to achieve very high cell densities sustainably is the cascade system developed in the Czech Republic and used for cultivation of *Chlorella* (Setlik et al., 1970). With a culture depth of less than 1 cm, cell densities of up to

TABLE 5.2
Examples of Productivities Achieved in Open Ponds with Various Microalgal Species

Species	Highest Productivity		Ref.
	g m^{-2}d^{-1}	g l^{-1}d^{-1}	
Amphora	39		Sheehan et al., 1998
Ankistrodesmus falcatus		0.18	Sheehan et al., 1998
Chaetoceros muelleri	26	0.18	Sheehan et al., 1998
Chlorella sp.	14		Sheehan et al., 1998
Chlorella sp.	2		Setlik et al., 1970
Chlorella vulgaris	40		Piorreck et al., 1984
Cyclotella cryptica	27		Sheehan et al., 1998
Isochrysis galbana	28		Sheehan et al., 1998
Nannochloropsis	15		Sheehan et al., 1998
Nannochloropsis salina	25		Sheehan et al., 1998
Porphyridium purpureum		0.18	Sheehan et al., 1998
Scenedesmus obliquus	48		Grobbelaar, 2000
Spirulina platensis	12		Piorreck et al., 1984
Spirulina platensis	27	0.18	Richmond et al., 1990
Spirulina sp.	69		De Morais et al., 2009
Spirulina sp.	19	0.15, 0.32	Pushparaj et al., 1997
Tetraselmis suecica	19		Sheehan et al., 1998

10 g L^{-1} *Chlorella* were achieved. However, the biomass productivity was comparable to that of raceways (25 g $m^{-2}d^{-1}$) (Lee, 2001). The system had a sloping base made of glass, which rendered it very expensive, but the use of cheaper materials could make it price competitive with raceway ponds. A similar system has been used in Western Australia, consisting of a 0.5-ha sloping, plastic-lined pond for the production of *Chlorella*, achieving similar biomass productivity (Borowitzka, 1999).

5.3.2 CLOSED SYSTEMS

Although they are more expensive to build and run than open systems, the promise of improved yields, and the possibility of growing a wider range of species, has led to significant interest in closed reactors. It is much easier to control contamination and environmental parameters in closed systems, allowing the cultivation of more sensitive strains and expanding the potential product range. Biomass concentrations obtained are higher than in open systems, thus reducing the cost of harvesting. However, the capital and operating costs of closed reactors are higher than those of open ponds (Carvalho et al., 2006).

A large variety of PBR designs have been proposed, only a few of which have been commercialized (Greenwell et al., 2010). Most designs are based on the premise of optimizing light provision by maximizing the area-to-volume ratio, while ensuring a reasonable working volume, cost of reactor material, and mixing pattern (Carvalho et al., 2006). One of the major problems with closed reactors is temperature control, and the larger the area-to-volume ratio, the more susceptible the temperature of the medium is to changes in environmental temperature. The optimum light path length is 2 to 4 cm (Borowitzka, 1999), but most closed reactors have a larger diameter for ease of mixing, cleaning, temperature regulation, and to increase the working volume while reducing the cost of construction materials. Sedimentation is prevented by maintaining turbulent flow through mixing mechanically or by airlift.

An important and often overlooked feature of closed reactors is the ease with which they can be cleaned and sanitized (Chisti, 2007). Closed microalgal reactors are often presented as having the advantage of a decreased risk of contamination. Contamination can be avoided in closed reactors, but only if they are operated under sterile conditions, which adds to the cost (Scott et al., 2010). Due to their large size and surface area, closed reactors cannot be effectively sterilized by heat, and therfore require chemical sterilizers. These are not always 100% effective and sometimes require large volumes of sterile water to flush out the chemical agent. Most closed PBRs do not satisfy the good manufacturing practice requirements for production of pharmaceutical products (Lee, 2001).

The most common designs are tubular (Miyamoto et al., 1988; Richmond et al., 1993; Borowitzka, 1996; Vonshak, 1997) or flat-plate (Hu et al., 1996; Vonshak, 1997) reactors. Both types usually operate with culture circulated between a light-harvesting unit, consisting of narrow tubes or plates, to provide a high surface area, and a reservoir or gas exchange unit in which CO_2 is supplied, O_2 removed, and harvesting carried out. The circulating pump must be carefully designed so as to avoid shear forces disrupting algal cells. A variety of microalgae, including *Chlorella* and *Spirulina*, have been successfully maintained in both tubular and flat-plate PBRs (Molina Grima, 1999; Lee, 2001; Pulz, 2001; Carvalho et al., 2006).

Scale-up of any PBR design is challenging due to the difficulty in maintaining optimum light, temperature, mixing, and mass transfer in large volumes (Ugwu et al., 2008). Large-scale closed systems will likely be based on the integration of multiple units rather than increasing the size of a single reactor (Brennan and Owende, 2010).

5.3.2.1 Tubular Reactors

Tubular reactors are characterized by very high area-to-volume ratios (dependent on tube diameter) but poor mass transfer, leading to O_2 build-up and CO_2 depletion (resulting in photorespiration, oxidative damage, and pH gradients) over the length of closed tubing. Tubes are generally manufactured from polyethylene or glass. The most important criteria for construction material are transparency, to allow good light penetration, and low cost (Ugwu et al., 2008). Additional challenges include photo-inhibition, temperature control, and fouling due to cells adhering to the inside of tube walls, leading to decreased light penetration (Ugwu et al., 2008). Narrow-diameter tubes can present a challenge to clean.

Most tubular reactors can be categorized into one of three types:

1. Vertical airlifts or bubble columns consisting of a clear vertical tube mixed by gas sparging from the bottom
2. Horizontal tubular systems with clear, thin-diameter tubing lying or stacked horizontally, usually connected to a gas transfer system
3. Helical tubular reactors consisting of thin, flexible tubing coiled around a circular framework

Airlift and bubble column reactors (Figures 5.2a and b) are examples of vertical tubular reactors. Air, or air enriched with CO_2, is bubbled into the bottom, providing efficient mixing and gas transfer throughout the reactor. The simplest form of bubble column reactor is a hanging polyethylene bag, and these have frequently been used as a low-cost option. Plastic bags have a high transparency, good sterility at start-up (due to the high temperatures used in plastic extrusion), and are readily replaceable. Concentrations three times that of open ponds were obtained by culturing *Porphyridium* in 25 L hanging bags (Cohen and Arad, 1989). Other researchers have also found that 40 to 50 L bags are practical (Trotta, 1981; Martínez-Jerónimo and Espinosa-Chávez, 1994).

Although cultivation in plastic bags is simple, cheap, and widely employed, particularly in the production of microalgae as feed for aquaculture hatcheries, scale-up is limited by the fragility of cheap plastic and light penetration, as increases in bag volume lead to decreased productivity due to mutual shading (Martínez-Jerónimo and Espinosa-Chávez, 1994). Rigid vertical tubes have also been frequently used (Carvalho et al., 2006). In an airlift reactor, an inner tube called the riser directs air bubbles up the center of the reactor and then down the outer region, called the downcomer (Figure 5.2b). This provides effective, gentle mixing and produces regular light-dark cycles.

Vertical reactors are compact, low cost, and easy to clean and keep sterile (Ugwu et al., 2008); however, their size is limited by the surface-to-volume ratio. The scale-up of any tank, container, or hanging bag becomes limited by light penetration at a

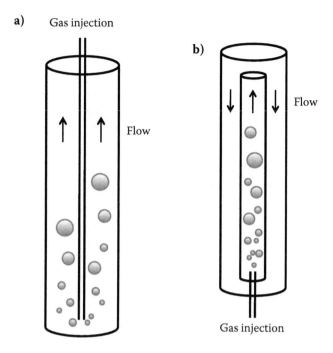

FIGURE 5.2 Schematic diagram of (a) vertical bubble column and (b) airlift reactor.

volume of between 50 and 100 L. Underwater lighting has been considered, either in the form of submersible lamps, or optical fibers redirecting sunlight, but these add to the cost and energy footprint of the system (Pulz, 2001). In addition, the vertical angle of reactors is not ideal for capturing incident sunlight (Carvalho et al., 2006).

Horizontal tubular reactors are designed to optimize light capture by increasing the angle to sunlight (Figure 5.3). Internal tube diameters range between 1.2 and 13 cm (Lee, 2001). The thinner the tubing used in the solar collector, the more efficient the light capture but the lower the culture volume per length of tubing. Thin tubing is also particularly susceptible to overheating, and temperature regulation mechanisms must be installed—for example, evaporative water cooling (often requiring large volumes of water), immersing the tubes in water, or shading them by covering or overlapping the tubes. The length of the closed tubing is constrained by O_2 build-up; therefore, reactors are usually modular, with parallel sets of shorter tubes interconnected, rather than a single long tube. Gas transfer takes place either at tube connections or in a dedicated gas exchange unit, where aeration and mixing are provided either by pump or airlift (Ugwu et al., 2008). There have been successful runs of horizontal tubular reactors with volumes of up to 8,000 to 10,000 L (Torzillo et al., 1986). One of the major disadvantages of horizontal reactors is the large land area required for horizontal tubing. The increased productivity with respect to open ponds may not be cost effective if a greater land area is necessary (Carvalho et al., 2006).

Helical tubular reactors (Figure 5.4) are a promising alternative to horizontal tubular reactors as they reduce the land area required. By expanding vertically,

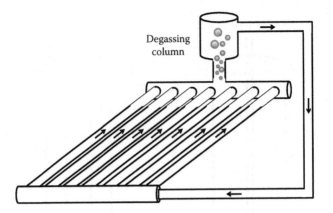

FIGURE 5.3   Schematic diagram of a horizontal tubular reactor.

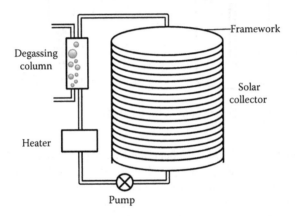

FIGURE 5.4   Schematic diagram of a helical tubular reactor.

the areal footprint of the reactor is smaller, but the angle to sunlight is also reduced. Placing a light in the center of the coil can improve light penetration. A conical, instead of helical, framework has also been suggested, as it improves the spatial distribution of tubes for sunlight capture (Morita et al., 2000). However, scale-up is then limited as the angle and height of the coil are defined. One of the most effective designs is the Biocoil developed by Robinson (1987). It consists of a set of polyethylene plastic tubes (2.4 to 5 cm in diameter) wound helically around an open circular framework. Parallel bands of tubes connect to a gas exchange tower. A centrifugal pump is used to move culture to the top of the coil, which may not be suitable for all species due to shear stress in sensitive cells and pump fouling in filamentous species. A heat exchanger or evaporative cooling provides temperature control. The system provides good mixing; minimal cell adhesion to the inside of tubes and scale-up is easy, involving the addition of more parallel coils. Several marine species and *Spirulina* have been successfully cultivated for more than 4 months in a 700 L Biocoil (Borowitzka, 1999; Carvalho et al., 2006).

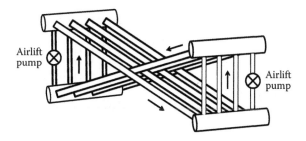

FIGURE 5.5   Schematic diagram of an α-shaped tubular reactor.

An α-shaped reactor, presenting an interesting alternative tube layout, was designed by Lee et al. (1995) (Figure 5.5). The symmetrical design uses airlift pumps to aerate and mix the culture in vertical tubes at either end. This then flows down tubes at a 45° angle, thus maximizing the angle of tubes to sunlight while saving space. As flow through the entire system is in the same direction, with two CO_2 injection points, relatively high liquid flow and mass transfer rates can be maintained with relatively low air supply rates (Lee et al., 1995).

5.3.2.2   Flat-Plate Reactors

Flat-plate reactors are characterized by a large surface area and lower O_2 accumulation than tubular reactors (Ugwu et al., 2008). They generally consist of narrow panels, with walls made of glass or stiff Perspex® (Figure 5.6a). Productivity is maximal at minimum light path length, but again the increased yield must be traded off against increased cost of materials to hold the same volume of culture. Reactors are usually modular, with working volumes of up to 1,000 L (Carvalho et al., 2006), and can be set up vertically or at an angle to the horizontal (Lee, 2001). The panels can have an open headspace for improved gas transfer, although such an open zone can compromise sterility. They are normally cooled either by spraying the flat surface with water (which can be collected for reuse by a trough at the base of the panel), or by sandwiching two panels together (one for algal growth and one for temperature modulation) (Tredici et al., 1991). In the past, there have been problems with circulation in flat-plate reactors (Carvalho et al., 2006), particularly at the base and in the corners of square panels. The main advantages of flat-plate reactors are their uniform light distribution, the fact that reactors can be tilted to maintain optimal orientation toward the sun, and a reduced need for pumping if the culture is mixed by air.

A promising modification to the basic flat-plate design is the alveolar panel (Figure 5.6b). Alveolar reactors have flat panels divided into a series of internal channels (or alveoli) providing structural rigidity and enabling efficient flow of culture medium (Greenwell et al., 2010). The walls are made of polycarbonate, PVC, or polymethyl methacrylate (Carvalho et al. 2006). Tredici et al. (1991) used double sets of alveolar plates, placed horizontally, with culture circulated in the upper set and the lower set acting as a thermostat to control the temperature. Alveolar plates have also been placed vertically, with air bubbling from the bottom of each channel. A comparison of the productivities achieved in a range of closed PBRs is presented in Table 5.3.

a) 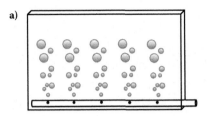   b)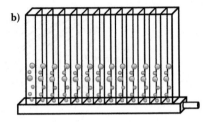

FIGURE 5.6   Schematic diagram of (a) flat plate and (b) alveolar panel reactor.

5.3.3 ALTERNATIVE DESIGNS

5.3.3.1 Stirred Tank Fermenter

Conventional heterotrophic fermenters (used routinely for cultivation of nonphotosynthetic microorganisms) have been used for the production of microalgae, particularly for high-value products such as fine chemicals and pharmaceuticals (Mata et al., 2010). The area-to-volume ratio of a stirred tank is low; therefore, some form of internal illumination (e.g., artificial lighting or sunshine directed through optical fibers) is necessary, or cultures must be grown heterotrophically (Lee, 2001; Carvalho et al., 2006). Some algae are able to grow mixotrophically or heterotrophically on organic substrates such as glucose, acetate, or peptone (Grobbelaar, 2009). In this case, part or all of the carbon and energy is supplied by the organic substrate, thereby reducing the dependence of growth rate on light and CO_2 provision. Mixotrophic growth rates (where cells utilize both light and organic substrates) are often greater than purely photo-autotrophic or heterotrophic (e.g., *Chlorella* and *Haematococcus*) (Lee, 2001).

The main advantages of stirred tank reactors are the precise control over operating parameters, the ability to maintain sterility, and the wealth of experience in their operation and scale-up with yeast and microbes that exists. Maintaining sterility of cultures is crucial for the production of certain high-value metabolites (e.g., pharmaceuticals). *Chlorella* is routinely grown in stirred tanks up to high cell density (45 g L^{-1}), with a volumetric productivity of up to 20 g $L^{-1}d^{-1}$ (Lee, 2001). When an organic substrate is added to the medium, sterility becomes a priority as bacteria readily compete with algae for the dissolved nutrients (Lee, 2001). Stirred tanks of up to 250 L have been run (Carvalho et al. 2006). Ogbonna et al. (1999) investigated the use of stirred tanks with a combination of sunlight and internal artificial lighting, which may reduce costs. Cultivation in stirred tank systems is limited to species able to assimilate organic carbon substrates. Not all algae are able to grow heterotrophically (Lee, 2001).

5.3.3.2 Wave/Oscillatory Flow Reactors

Oscillatory flow bioreactors contain equally spaced orifice plate baffles in a tubular style reactor. This reactor has improved heat and mass transfer due to the oscillatory motion that is imposed on the net flow of the fluid, which means that the degree of mixing is independent of the net flow. This results in long residence times in relatively low length-to-diameter ratios. This reactor design, therefore, has the potential to decrease the energy required for mixing algae cultures, due to decreased pumping requirements, and also leads to decreased capital costs (Harvey et al., 2003).

TABLE 5.3

Examples of Productivities Achieved in Closed Bioreactors with Various Microalgal Species

Reactor Type	Species	Highest Productivity		Ref.
		$g\ m^{-2}d^{-1}$	$g\ l^{-1}d^{-1}$	
Vertical column				
	Phaeodactylum		0.69	Sánchez Mirón et al., 1999
	Isochrysis galbana		1.60	Qiang and Richmond, 1994
	Tetraselmis	38.2	0.42	Chini Zittelli et al., 2006
	Haematococcus pluvialis		0.06	López et al., 2006
Airlift tubular				
	Haematococcus pluvialis		0.41	López et al., 2006
	Porphyridium cruentum		1.5	Rubio et al., 1999
	Porphyridium cruentum	20	1.2	Acién Fernández et al., 2001
	Phaeodactylum	32	1.9	Molina Grima et al., 2001
Horizontal tubular				
	Spirulina maxima	25	0.25	Torzillo et al., 1986
	Spirulina sp.	27.8		Torzillo et al., 1986
	Spirulina platensis	27	1.60	Richmond et al., 1993
	Isochrysis galbana		0.32	Molina Grima et al., 1994
	Phaeodactylum		2.02	Fernandez et al., 1998
	Phaeodactylum		2.76	Grima et al., 1996
	Haematococcus pluvialis	13	0.05	Olaizola, 2000
Inclined tubular				
	Chlorella sp.	72.5	2.90	Lee et al., 1995
	Chlorella sp.	130	3.64	Lee and Low, 1991
	Chlorella sorokiniana		1.47	Ugwu et al., 2002
Helical tubular				
	Phaeodactylum tricornutum		1.4	Hall et al., 2003
	Tetraselmis chuii		1.20	Borowitzka, 1997
Flat plate				
	Spirulina platensis	33	0.30	Hu et al., 1996
	Spirulina platensis	51	4.30	Hu et al., 1996
	Spirulina platensis	24	0.80	Tredici et al., 1991
	Nannochloropsis		0.27	Cheng-Wu et al., 2001
	Nannochloropsis		0.85	Richmond and Cheng-Wu, 2001
	Haematococcus pluvialis	10.2		Huntley and Redalje, 2006
	Chlorella sp.	22.8, 19.4	3.8, 3.2	Doucha et al., 2005

5.3.3.3   Hybrid Production Systems

As open and closed systems offer different advantages and disadvantages, it seems practical that a combination of the two could provide the best of both worlds. This idea has been investigated in various configurations, either by circulating culture between open and closed reactors through a single growth stage, or by having a two-stage culture regime where cells are moved from one to the other at a certain point. A simple single-stage intermediate design is produced by enclosing or semi-enclosing open ponds in tunnels or greenhouses to improve temperature control and reduce evaporation and contamination. This is very effective in improving productivity, particularly in colder seasons, but comes at increased capital cost (Vonshak, 1997). Pushparaj et al. (1997) described a system where an alveolar panel oriented toward the sun was coupled with an open raceway for gas transfer. Adding the panels improved the productivity of the pond from 0.18 to 0.31 g $L^{-1}d^{-1}$.

In two-stage configurations, culture is usually grown initially in closed PBRs to optimize the growth rate and minimize contamination of the inoculum, which is then moved to an open pond for the second growth stage. Integration of open and closed PBRs in this way could provide sufficiently large, clean inoculants to limited-duration culture in outdoor raceways in order to significantly limit adverse events (Greenwell et al., 2010).

The second cultivation stage often involves nutrient stress for accumulation of a metabolite such as lipids or pigments. The nutrient stress stage is suited to open ponds because the growth rate is naturally low and therefore not affected by the low light availability (Brennan and Owende, 2010). Initial culture in closed reactors also implies that the culture entering the pond is relatively dense and therefore less likely to be contaminated, particularly in a nutrient-deprived environment (Singh et al., 2011). This sort of system has been used for the production of astaxanthin from *Haematococcus* (Huntley and Redalje, 2006) and described for the production of biodiesel from *Nannochloropsis* (Rodolfi et al., 2009).

5.4   COMPARISON OF REACTOR TYPES

The microalgal reactors described above differ in features such as surface-to-volume ratio, freedom to adjust orientation and inclination, efficiency of mixing and gas supply (related to hydrodynamics and mass transfer), ease of maintenance, temperature regulation, and construction materials. Table 5.4 presents a comparison of these design features in six major types of reactor. No reactor design is able to effectively control all these parameters simultaneously; therefore, any choice will be a compromise between the advantages and disadvantages of each system (Table 5.5).

5.4.1   The Open versus Closed System Debate

The relative merits of closed and open systems have been extensively debated in the microalgal literature (Pulz, 2001; Carvalho et al., 2006; Grobbelaar, 2009; Mata et al., 2010). There is no doubt that open ponds are the primary systems used in large-scale, outdoor microalgal cultures, but their commercial use has been limited to species that can be maintained using an extreme cultivation environment

TABLE 5.4
Comparison of Main Design Features of Various Reactor Types

Reactor Type	Raceway	Vertical Column or Airlift	Horizontal Tubular	Helical Tubular	Flat Plate	Stirred Tank
Mixing	Fair	Good	Uniform	Uniform	Largely uniform	Largely uniform
Gas transfer	Poor	Good	Low to high along length of tube	Low to high along length of tube	Good	Low to high
Hydrodynamic stress	Low	Low	Low to high depending on pumping system	Low to high depending on pumping system	Low	High
Light harvesting efficiency (area:volume ratio)	Fair, depending on depth	Good at low volume; poor at high volume	Excellent at narrow diameter	Excellent at narrow diameter	Excellent at narrow diameter	Poor
Temperature control	None	Good	Good	Good	Good	Good
Species control	Difficult	Easy	Easy	Easy	Easy	Easy
Sterility	None	Easily achievable	Potentially achievable	Potentially achievable	Potentially achievable	Easily achievable
Land area required	Large	Small	Large	Medium	Small	Small

Source: From Borowitzka (1999); Carvalho et al. (2006).

TABLE 5.5
Advantages and Disadvantages of Different Reactor Types

Reactor Type	Advantages	Disadvantages
Raceway	Relatively economical Low energy input Easy to clean and maintain No O_2 build-up Large production capacity	Little control of culture conditions Poor mixing Light limitation Readily contaminated Limited growing period Poor productivity Large land area required Limited CO_2 mass transfer with large CO_2 losses to atmosphere if pond sparged Temperature determined by climate
Vertical column PBR	Excellent mixing High mass transfer rates Low shear stress Easy to clean and sterilize Reduced photo-inhibition Low cost Low land area requirement	Low angle to incident sunlight Size limited by area:volume ratio Large mixing energy due to gas compression
Tubular PBR	Large illuminated surface area Good biomass productivity Relatively cheap construction materials	Gradients of pH, O_2, and CO_2 along tubes Wall growth: fouling in tubes difficult to clean Large land area required Significant water losses if evaporative cooling is used Possible hydrodynamic stress
Flat-plate PBR	Large illuminated surface area Short light path Good biomass productivity Easier to clean Lower O_2 build-up	Scale-up requires many modules – material intensive Temperature control critical in thin reactors

Source: From Borowitzka (1999); Pulz (2001); Chisti (2007); Ugwu et al. (2008); Brennan and Owende (2010); Mata et al. (2010).

(Lee, 2001). To expand the product range, there is significant interest in the design of closed reactors, particularly in the production of high-value, low-volume products requiring a high degree of sterility. The essence of the debate is presented in Table 5.6 through a comparison of the key parameters of open and closed reactors.

Despite their higher cost and technical complexity, closed systems promise great improvements in enhancing control over process parameters. The challenge appears to

TABLE 5.6

Comparison of Key Design Features and Process Parameters of Open and Closed Systems

	Open Systems	Closed Systems
Species Control		
Choice of species	Restricted	Flexible
Main criteria for species selection	Growth competition	Shear resistant
	Tolerance of range of conditions	Temperature tolerant
		O_2 resistant
Contamination risk	High	Reduced
Sterility	None	Achievable
Light Availability		
Light utilization efficiency	Low	High
Area-to-volume ratio	Low (5–10 m^{-1})	High (20–200 m^{-1})
Process Control		
Mixing	Poor	Uniform
Gas transfer	Poor	Fair/high
CO_2 loss	High	Depends on pH, alkalinity, gas recycling
O_2 build-up	Low	High
Overheating problems	Low	High
Temperature control	Harder, but cooling not as necessary due to large volume	Easier, but cooling more often necessary
Hydrodynamic stress	Low	High
Evaporative water loss	Surface evaporation	Depends on cooling and sparging design
Weather dependence	High	Less
Cultivation period	Limited	Extended
Productivity		
Biomass concentration	Low (<1 g L^{-1})	High (>2 g L^{-1})
Biomass productivity	Low	High
Reproducibility of production	Variable but consistent over time	Possible within certain tolerances
Cost		
Capital cost	Low	High
Most costly operating parameters	Mixing	Temperature and oxygen control
Energy input required	Low	High
Harvesting efficiency	Low due to low biomass concentration	Higher due to high biomass concentration

Source: From Pulz (2001); Carvalho et al. (2006); Grobbelaar (2009); Mata et al. (2010).

lie in enhancing productivity sufficiently that it outweighs the additional cost of closed reactors. Another alternative is to attempt to design PBRs that are cheap to build in terms of construction materials, as well as efficient in terms of light distribution, mixing, gas sparging, etc., which makes them cheap to operate by lowering energy requirements.

A major but rarely recognized concern, particularly for energy products such as biofuels, is the energy balance of the production system. For a process to be economically viable and sustainable, the energy generated when the product is used must be greater than that involved in its manufacture. The energy inputs in microalgal reactors are particularly focused on the mixing and gas pressurization, as well as the embodied energy in reactor materials; therefore, open systems have a more favorable energy balance than closed systems (Richardson, 2011).

5.5   CONCLUSION

In the production of algal energy products, the aim is the biological conversion of sunlight to a more convenient, portable, storable, and accessible form of fuel. In the case of biodiesel production, this entails the production of algal lipids. Lipid productivity is dependent on both biomass productivity and lipid content (Griffiths and Harrison, 2009), which is determined by both the species used and the culture conditions provided by the reactor.

Most large-scale commercial algal production systems to date have been for food, feed, neutraceutical, or fine chemical production. As biofuel is a bulk commodity product, production must be on a grand scale, and costs must be extremely low. Sterility, particularly microbial contamination, is perhaps less of a concern for energy production than it would be for a product such as a neutraceutical or fine chemical for human consumption. A particular consideration with an energy product is that the energy balance must be positive; that is, the energy recovered from the product must exceed the energy input required for production. LCA (life cycle assessment) studies to date suggest that biofuel production in closed reactors is unable to achieve a net energy ratio (energy out/energy into process) of above one (Lardon et al., 2009; Richardson, 2011).

It is generally considered that closed PBRs alone will be incapable of cost-effectively producing microalgal biomass on the scale required for biofuels (Greenwell et al., 2010). While productivities will inevitably be lower in open raceways, it is envisaged that open systems, due to their lower cost, simplicity of operation, and ability to scale to large volumes, will form the basis of microalgal production for biofuels (Sheehan et al., 1998). The lipids necessary for biodiesel production are often produced under nutrient stress conditions. Therefore, it is likely that a two-phase system using closed reactors to generate contamination-free inoculum with a high biomass concentration for a second product-generating stage in open systems could be advantageous.

REFERENCES

Acién Fernández, F.G., Fernández Sevilla, J.M., Sánchez Pérez, J.A., Molina Grima, E., and Chisti, Y. (2001). Airlift-driven external-loop tubular photobioreactors for outdoor production of microalgae: Assessment of design and performance. *Chemical Engineering Science*, 56: 2721–2732.

Borowitzka, M. (1996). Closed algal photobioreactors: Design considerations for large-scale systems. *Journal of Marine Biotechnology*, 4: 185–191.

Borowitzka, M. (1999). Commercial production of microalgae: Ponds, tanks, and fermenters. *Progress in Industrial Microbiology*, 70: 313–321.

Borowitzka, M.A. (1997). Microalgae for aquaculture—Opportunities and constraints. *Journal of Applied Phycology*, 9: 393–401.

Brennan, L., and Owende, P. (2010). Biofuels from microalgae—A review of technologies for production, processing, and extractions of biofuels and co-products. *Renewable and Sustainable Energy Reviews*, 14: 557–577.

Carvalho, A., Meireles, L., and Malcata, F. (2006). Microalgal reactors: A review of enclosed system designs and performances. *Biotechnology Progress*, 22: 1490–1506.

Cheng-Wu, Z., Zmora, O., Kopel, R., and Richmond, A. (2001). An industrial-size flat plate glass reactor for mass production of *Nannochloropsis* sp. (Eustigmatophyceae). *Aquaculture*, 195: 35–49.

Chini Zittelli, G., Rodolfi, L., Biondi, N., and Tredici, M.R. (2006). Productivity and photosynthetic efficiency of outdoor cultures of *Tetraselmis suecica* in annular columns. *Aquaculture*, 261: 932–943.

Chisti, Y. (2007). Biodiesel from microalgae. *Biotechnology Advances*, 25: 294–306.

Chisti, Y. (2008). Biodiesel from microalgae beats bioethanol. *Trends in Biotechnology*, 26: 126–131.

Cohen, E., and Arad, S. (1989). A closed system for outdoor cultivation of *Porphyridium*. *Biomass*, 18: 59–67.

De Morais, M.G., Radmann, E., Andrade, M., Teixeira, G., Brusch, L., and Costa, J.A.V. (2009). Pilot scale semicontinuous production of *Spirulina* biomass in southern Brazil. *Aquaculture*, 294: 60–64.

Dennis, D., Turpin, D., Lefebvre, D., and Layzell, D. (1998). *Plant Metabolism, 2nd ed.*, Singapore: Longman.

Doucha, J., Straka, F., and Livansky, K. (2005). Utilization of flue gas for cultivation of microalgae *Chlorella* sp. in an outdoor open thin-layer photobioreactor. *Journal of Applied Phycology*, 17: 403–412.

Fernandez, F., Camacho, F., Perez, J., Sevilla, J., and Grima, E. (1998). Modeling of biomass productivity in tubular photobioreactors for microalgal cultures: Effects of dilution rate, tube diameter, and solar irradiance. *Biotechnology and Bioengineering*, 58: 605–616.

Greenwell, H.C., Laurens, L.M.L., Shields, R.J., Lovitt, R.W., and Flynn, K.J. (2010). Placing microalgae on the biofuels priority list: A review of the technological challenges. *Journal of the Royal Society, Interface/The Royal Society*, 7: 703–726.

Griffiths, M.J., and Harrison, S.T.L. (2009). Lipid productivity as a key characteristic for choosing algal species for biodiesel production. *Journal of Applied Phycology*, 21: 493–507.

Grima, E.M., Pérez, J.A.S., Camacho, F.G., Sevilla, J.M.F., and Fernández, F.G.A. (1996). Productivity analysis of outdoor chemostat culture in tubular air-lift photobioreactors. *Journal of Applied Phycology*, 8: 369–380.

Grobbelaar, J.U. (2000). Physiological and technological considerations for optimising mass algal cultures. *Journal of Applied Phycology*, 12: 201–206.

Grobbelaar, J.U. (2009). Factors governing algal growth in photobioreactors: The "open" versus "closed" debate. *Journal of Applied Phycology*, 21: 489–492.

Hall, D.O, Fernández, F.G.A., Guerrero, E.C., Rao, K.K., and Grima, E.M. (2003). Outdoor helical tubular photobioreactors for microalgal production: Modeling of fluid-dynamics and mass transfer and assessment of biomass productivity. *Biotechnology and Bioengineering*, 82: 62–73.

Harvey, A.P, Mackley, M.R., and Seliger, T. (2003). Process intensification of biodiesel production using a continuous oscillatory flow reactor. *Journal of Chemical Technology & Biotechnology*, 78: 338–341.

Hu, Q., Guterman, H., and Richmond, A. (1996). A flat inclined modular photobioreactor for outdoor mass cultivation of photoautotrophs. *Biotechnology and Bioengineering*, 51: 51–60.

Huntley, M.E., and Redalje, D.G. (2006). CO_2 mitigation and renewable oil from photosynthetic microbes: A new appraisal. *Mitigation and Adaptation Strategies for Global Change*, 12: 573–608.

Kumar, A., Ergas, S., Yuan, X., Sahu, A., Zhang, Q., Dewulf, J., Malcata, X., and Langenhove, H.V. (2010). Enhanced CO_2 fixation and biofuel production via microalgae: Recent developments and future directions. *Trends in Biotechnology*, 28: 371–380.

Lardon, L., Hélias, A., Sialve, B., Steyer, J.P., and Bernard, O. (2009). Life-cycle assessment of biodiesel production from microalgae. *Environmental Science & Technology*, 43(17): 6475–6481.

Lee, Y.K. (2001). Microalgal mass culture systems and methods: Their limitation and potential. *Journal of Applied Phycology*, 13: 307–315.

Lee, Y.K., and Low, C.S. (1991). Effect of photobioreactor inclination on the biomass productivity of an outdoor algal culture. *Biotechnology and Bioengineering*, 38: 995–1000.

Lee, Y.K., Ding, S.Y., Low, C.S., Chang, Y.C., Forday, W.L., and Chew, P.C. (1995). Design and performance of an α-type tubular photobioreactor for mass cultivation of microalgae. *Journal of Applied Phycology*, 7: 47–51.

López, M.C.G.M, Sánchez, E.D.R., López, J.L.C., Fernández, F.G.A., Sevilla, J.M.F., Rivas, J., Guerrero, M.G., and Grima, E.M. (2006). Comparative analysis of the outdoor culture of *Haematococcus pluvialis* in tubular and bubble column photobioreactors. *Journal of Biotechnology*, 123: 329–342.

Martinez-Jeronimo, F., and Espinosa-Chavez, F. (1994). A laboratory-scale system for mass culture of freshwater microalgae in polyethylene bags. *Journal of Applied Phycology*, 6: 423–425.

Mata, T.M., Martins, A., and Caetano, N.S. (2010). Microalgae for biodiesel production and other applications: A review. *Renewable and Sustainable Energy Reviews*, 14: 217–232.

Miyamoto, K., Wable, O., and Benemann, J.R. (1988). Vertical tubular reactor for microalgae cultivation. *Biotechnology Letters*, 10: 703–708.

Molina Grima, E. (1999). Photobioreactors: light regime, mass transfer, and scaleup. *Journal of Biotechnology*, 70: 231–247.

Molina Grima, E., Fernandez, J., Acien Fernandez, F., and Chisti, Y. (2001). Tubular photobioreactor design for algal cultures. *Journal of Biotechnology*, 92: 113–131.

Molina Grima, E., Sánchez Pérez, J.A., García Camacho, F., García Sánchez, J.L., Acién Fernández, F.G., and López Alonso, D. (1994). Outdoor culture of *Isochrysis galbana* ALII-4 in a closed tubular photobioreactor. *Journal of Biotechnology*, 37: 159–166.

Morita, M., Watanabe, Y., and Saiki, H. (2000). Investigation of photobioreactor design for enhancing the photosynthetic productivity of microalgae. *Biotechnology and Bioengineering*, 69: 693–698.

Ogbonna, J.C, Soejima, T., and Tanaka, H. (1999). An integrated solar and artificial light system for internal illumination of photobioreactors. *Journal of Biotechnology*, 70: 289–297.

Olaizola, M. (2000). Commercial production of astaxanthin from *Haematococcus pluvialis* using 25,000-liter outdoor photobioreactors. *Journal of Applied Phycology*, 12: 180–189.

Piorreck, M., Baasch, K.H., and Pohl, P. (1984). Biomass production, total protein, chlorophylls, lipids and fatty acids of freshwater green and blue-green algae under different nitrogen regimes. *Phytochemistry*, 23: 207–216.

Pulz, O. (2001). Photobioreactors: Production systems for phototrophic microorganisms. *Applied Microbiology and Biotechnology*, 57: 287–293.

Pushparaj, B., Pelosi, E., Tredici, M., Pinzani, E., and Materassi, R. (1997). An integrated culture system for outdoor production of microalgae and cyanobacteria. *Journal of Applied Phycology*, 9: 113–119.

Qiang, H., and Richmond, A. (1994). Optimizing the population density in *Isochrysis galbana* grown outdoors in a glass column photobioreactor. *Journal of Applied Phycology*, 6: 391–396.

Richardson, C. (2011). Investigating the Role of Reactor Design to Maximise the Environmental Benefit of Algal Oil for Biodiesel. Master's thesis. University of Cape Town, South Africa.

Richmond, A. (2000). Microalgal biotechnology at the turn of the millennium: A personal view. *Journal of Applied Phycology*, 12: 441–451.

Richmond, A., and Cheng-Wu, Z. (2001). Optimization of a flat plate glass reactor for mass production of *Nannochloropsis* sp. outdoors. *Journal of Biotechnology*, 85: 259–269.

Richmond, A., Boussiba, S., Vonshak, A., and Kopel, R. (1993). A new tubular reactor for mass production of microalgae outdoors. *Journal of Applied Phycology*, 5: 327–332.

Richmond, A., Lichtenberg, E., Stahl, B., and Vonshak, A. (1990). Quantitative assessment of the major limitations on productivity of *Spirulina platensis* in open raceways. *Journal of Applied Phycology*, 2: 195–206.

Robinson, L. (1987). Improvements Relating to Biomass Production. European Patent EP0239272.

Rodolfi, L., Chini Zittelli, G., Bassi, N., Padovani, G., Biondi, N., Bonini, G., and Tredici, M.R. (2009). Microalgae for oil: Strain selection, induction of lipid synthesis and outdoor mass cultivation in a low-cost photobioreactor. *Biotechnology and Bioengineering*, 102: 100–112.

Rubio, F., Fernandez, F., Perez, J., Camacho, F., and Grima, E. (1999). Prediction of dissolved oxygen and carbon dioxide concentration profiles in tubular photobioreactors for microalgal culture. *Biotechnology and Bioengineering*, 62: 71–86.

Scott, S.A., Davey, M.P., Dennis, J.S., Horst, I., Howe, C.J., Lea-Smith, D.J., and Smith, A.G. (2010). Biodiesel from algae: challenges and prospects. *Current Opinions in Biotechnology*, 21: 277–286.

Setlik, I., Veladimir, S., and Malek, I. (1970). Dual purpose open circulation units for large scale culture of algae in temperate zones. I. Basic design considerations and scheme of a pilot plant. *Algologie Studies (Trebon)*, 1: 111–164.

Sheehan, J., Dunahay, T., Benemann, J.R., and Roessler, P. (1998). A Look Back at the US Department of Energy's Aquatic Species Program: Biodiesel from Algae. National Renewable Energy Laboratory, Golden, CO.

Singh, A., Nigam, P.S., and Murphy, J.D. (2011). Mechanism and challenges in commercialisation of algal biofuels. *Bioresource Technology*, 102: 26–34.

Spolaore, P., Joannis-Cassan, C., Duran, E., and Isambert, A. (2006). Commercial applications of microalgae. *Journal of Bioscience and Bioengineering*, 101: 87–96.

Sánchez Mirón, A., Contreras Gómez, A., García Camacho, F., Molina Grima, E., and Chisti, Y. (1999). Comparative evaluation of compact photobioreactors for large-scale monoculture of microalgae. *Journal of Biotechnology*, 70: 249–270.

Torzillo, G., Pushparaj, B., Bocci, F., Balloni, W., Materassi, R., and Florenzano, G. (1986). Production of *Spirulina* biomass in closed photobioreactors. *Biomass*, 11: 61–74.

Tredici, M.R., Carlozzi, P., Chini Zittelli, G., and Materassi, R. (1991). A vertical alveolar panel (VAP) for outdoor mass cultivation of microalgae and cyanobacteria. *Bioresource Technology*, 38: 153–159.

Trotta, P. (1981). A simple and inexpensive system for continuous monoxenic mass culture of marine microalgae. *Aquaculture*, 22: 383–387.

Ugwu, C.U., Aoyagi, H., and Uchiyama, H. (2008). Photobioreactors for mass cultivation of algae. *Bioresource Technology*, 99: 4021–4028.

Ugwu, C.U., Ogbonna, J.C., and Tanaka, H. (2002). Improvement of mass transfer characteristics and productivities of inclined tubular photobioreactors by installation of internal static mixers. *Applied Microbiology and Biotechnology*, 58: 600–607.

Vonshak, A. (1997). *Spirulina platensis* (Arthrospira): Physiology, Cell-Biology and Biotechnology, London: Taylor & Francis.

6 Harvesting of Microalgal Biomass

Manjinder Singh, Rekha Shukla, and Keshav Das
Biorefining and Carbon Cycling Program
College of Engineering
The University of Georgia
Athens, Georgia

CONTENTS

6.1   INTRODUCTION

Microalgae have been identified as a potential alternative resource for biofuel production. Significant drawbacks to algaculture include dilute culture density and the small size of microalgae, which translates into the need to handle large volumes of culture during harvesting. This energy-intensive process is therefore considered a major challenge for the commercial-scale production of algal biofuels. Most of the currently used harvesting techniques have several drawbacks, such as high cost, flocculant toxicity, or nonfeasibility of scale-up, which impact the cost and quality of products. As harvesting cost may itself contribute up to one-third of the biomass production cost, substantial amounts of research and development initiatives are needed to develop a cost- and energy-effective process for the dewatering of algae. Several factors, such as algae species, ionic strength of culture media, recycling of filtrate, and final products, should be considered when selecting a suitable harvesting technique. Harvesting cost and energy requirements must be reduced by a factor of at least 2 if algal biomass production is to be viable for very low-cost products such as biofuels. There could be considerable cost and energy savings in custom-designed, multi-stage harvesting techniques for algae farms. In such systems, a variety of harvesting technologies are arranged in a sequence based on culture chemistry, specific characteristics of each technique, and its energy requirements to dewater pond water to either 5% or 10–20% solids.

Techniques and processes of microalgae cultivation, harvesting, and dewatering have been reviewed extensively in the literature (Lee et al., 1998; Spolaore et al., 2006; Khan et al., 2009; Harun et al., 2010; Uduman et al., 2010). Due to the very dilute culture (<1.0 g of solids L^{-1}) and typically small size of microalgae with a diameter of 3 to 30 μm, large volumes must be handled to harvest algal biomass, which is an energy-intensive process. Therefore, harvesting microalgal biomass is considered a challenging issue for commercial-scale production of algal biofuels. Conventional processes used to harvest microalgae include concentration through centrifugation, foam fractionation, chemical flocculation, electro-flocculation, membrane filtration, and ultrasonic separation. The resulting high cost of biofuel production is a major bottleneck to its commercial application. The cost of harvesting may itself contribute to approximately 20% to 30% of the total cost of algal biomass, and the above methods would be viable only if the biomass harvested is used for extracting high-value products such as nutraceuticals (Girma et al., 2003). Harvesting, in general, can be defined as a series of processes for removing water from the algal growth culture and increasing the solids content from <1.0% to a consistency of up to 20% solids, depending on the downstream processing requirements for conversion to fuel. Thermal drying is generally discouraged (except when sufficient waste heat is available), because the amount of thermal energy needed to dry the algae would be a major fraction, if not all, of the energy content of the algal biomass.

Industrial production of algal biofuel is still in its infancy and therefore uncertainty in all stages of production and unpredictability of economy has been highly debated. Some very optimistic estimates on algae biofuels propose that the cost of algal oil production must be reduced by 5 to 6 times, in addition to the tax and environmental subsidies, to make them competitive with petroleum fuels (Chisti, 2007). For economical production of algal biomass, the selection of harvesting technologies is so crucial. Several factors, such as algal strain, ionic strength of culture media, recycling of filtrate, and final fate of harvested biomass, must be considered when selecting the harvesting technique. For example, the filamentous alga *Cladophora* with very long thread-like filaments (several centimeters long) lends itself to relatively cost-effective harvesting using membrane filtration. In contrast, chemical flocculation is not recommended if the harvested biomass must be processed for nutraceutical and pharmaceutical products because of the residual contamination caused by the flocculants. In general, it is quite difficult to recommend a single technique as the best for harvesting and recovery without consideration of specific process conditions and downstream product use. Scientists all over the world have developed several techniques for harvesting and recovery processes that rely on facts to simplify this overall process. Judicious exploitation of the different harvesting technologies is therefore necessary to reduce the harvesting cost and energy requirements by the desired factor of 2 if algal biomass production is targeted for very low-cost products such as biofuel. Advances in different methods of algal harvesting and dewatering to resolve our energy crisis, along with energy utilization by various techniques with respective constraints and drawbacks of each method, are discussed in this chapter. In addition, this chapter discusses the pros and cons of different algae harvesting techniques along with their energy requirements. The potential advantages of multi-stage hybrid harvesting systems involving more than one technique deployed in a specific sequence for efficient and energy-effective biomass recovery are discussed.

6.2 HARVESTING PROCESSES

The term *harvesting* refers to the concentration of dilute microalgal culture suspension to slurry or paste containing 5% to 25%, or more, total suspended solids (TSS). This slurry can be obtained in either a one-step or two-step harvesting process. Subsequent processing of the algal paste depends on the concentration of the algal paste. Increased product concentration decreases the cost of extraction and purification, as well as the effective unit cost of biomass. Concentration of algal paste significantly influences downstream processes, including drying. Microalgae are particles that have a colloidal character in suspension. Electric repulsive interaction between algal cells and cell interaction with the surrounding water provide stability to the algal suspension. Algal cells are usually characterized as negatively charged surfaces where the intensity of charge is a function of the species, ionic strength, and pH of the cultivation media (Taylor et al., 1998). These surface charges are helpful in the growth culture because they assist in keeping the cells in the water column so that they do not settle to the bottom of the pond, particularly in regions of the pond where the water velocity is low. However, the charges pose a challenge to the dewatering process because they eliminate the option of using a simple settling tank (or pond) for harvesting.

Harvesting and dewatering processes can be divided into two categories, namely (1) those in which the dewatering is performed directly on the algae culture, and (2) those involving agglomeration of the algae into macroscopic masses to facilitate the dewatering process. The former, which include centrifuges and membrane filters, avoid the complications and costs associated with the addition of coagulation and flocculation chemicals. Processes like flocculation, flotation and gravity sedimentation have acceptable energy requirements but have a fairly wide range of costs associated with motors and controls.

6.3 GRAVITY SEDIMENTATION

The sedimentation rates of algae are influenced by the settling velocity of microalgae, which can be increased by increasing cell dimensions (i.e., by aggregation of cells into large bodies) (Schenk et al., 2008). This principle is being applied to algal harvesting, wherein chemicals are added to enhance flocculation, causing the large algal flocs to settle more readily to the bottom of the container. The flocculation of algal biomass is generally followed by gravity sedimentation for settling of algal flocs, thus enhancing the efficiency of this process. Gravity sedimentation preceded by flocculation is one of the most commonly used techniques for first-stage (1% to 5% solids) algae biomass harvesting (Girma et al., 2003; Pittman et al., 2011). However, the gravity settling rate for very small sized microalgae is too low for routine high rate algae harvesting, and holding algal biomass for a long time under dark and static conditions can result in significant biomass loss via respiration and bacterial decomposition. Moreover, during flocculation, flocs may float due to adsorption of tiny air bubbles and do not settle by gravitational forces. The classical approach to gravity settling may therefore not be very efficient for rapid biomass recovery from high-rate algal ponds. The sedimentation rate can be increased by increasing the gravitational force via centrifugation. The latter has very high biomass recovery (>95%) and can be applied to a wide range of microalgae, although it cannot be used for an algae farm producing an energy feedstock, owing to cost constraints.

6.4   CENTRIFUGATION

Centrifugation is similar to sedimentation, wherein gravitational force is replaced by centrifugal acceleration to enhance the concentration of solids. Particle size and density difference are the key factors in centrifugal separation. Once separated, the algae concentrate can be obtained by simply draining the supernatant. Many researchers have advocated this method for reliable recovery of microalgae (Mohn, 1980; Benemann and Oswald, 1996; Girma et al., 2003). Different types of centrifuges have been used, and their respective reliability and efficiency have been documented by several researchers. For example, Heasman et al. (2000) reported that 90% to 100% harvesting efficiency can be achieved via centrifugation. Sim et al. (1988) compared centrifugation, chemical flocculation followed by dissolved air flotation (DAF), and membrane filtration processes for harvesting algae from pilot-scale ponds treating piggery wastewater, and they found that none of these processes were completely satisfactory. Centrifugation was reported to be very effective but too cost and energy intensive to be applied on a commercial scale. This kind of harvesting is usually recommended in the production of high-value metabolites or as a second-stage dewatering technique for concentrating algal slurries from 1% to 5% solids to >15% solids, as it does have some limitations. Undoubtedly, it is an efficient and reliable technique for microalgal recovery but one should also keep in mind its high operational cost.

6.5   FILTRATION

Filtration is the most competitive method compared to other harvesting techniques. It is most appropriately used for relatively large sized (>70 μm) algae such as filamentous species or agglomerates. Diatomaceous earth or cellulose can be used to increase filtration efficiency (Brennan and Owende, 2010). However, conventional filtration operated under pressure or suction is not suitable for smaller sized algae such as *Chlorella, Dunaliella,* and *Scenedesmus.* Membrane microfiltration and ultrafiltration are alternative filtration methods. The disadvantage of these processes is their high cost due to the frequent replacement of membranes and pumping costs (Pittman et al., 2011). There are many different types of filtration processes, such as dead-end filtration, microfiltration, ultrafiltration, pressure filtration, vacuum filtration, and tangential flow filtration (TFF). Mohn (1980) studied different pressure and vacuum filtration units for maximum dewatering of algae and warned against the use of these filters for harvesting *Coelastrum proboscideum* as he did not find it appropriate. However, he demonstrated that filtration processes can achieve a concentration factor of 245 times the original concentration for *C. proboscideum* to produce a slurry of 27% solids. Recent studies reveal that TFF and pressure filtration can be considered energy-efficient dewatering processes as they consume minimal amounts of energy considering the output and initial amounts of feedstock (Danquah et al., 2009). Simple filters can be used with centrifugation to achieve better results. Mohn (1980) and Danquah et al. (2009) have presented data on concentration factors and energy consumption of all filtration units. Large-scale recovery of microalgae using this technique is not recommended due to continuous fouling and the subsequent need to replace membranes. Few researchers have tried polymer membrane for continuous recovery. However, the performance of these membranes depends on several factors, such as hydrodynamic condition, concentration, and properties of microalgae. Although this method appears as an attractive dewatering method, the significant operating cost requirements cannot be overlooked.

6.6 FLOTATION

Floatation is a separation process based on the attachment of air bubbles to solid particles. The resulting flocs float to the liquid surface and are harvested by skimming and filtration. The success of flotation depends on the nature of suspended particles (microalgal cells in harvesting process). Air bubbles drift up the smaller particles (<500 μm) more easily (Matis et al., 1993). Also, the lower instability of suspended particles results in relatively higher air–particle contact. The attachment of air bubbles also depends on the air, solid, and aqueous phase contact angle. The larger the contact angle, the greater the tendency of air to adhere (Shelef et al., 1984). Dissolved air flotation (DAF), electrolytic flotation, and dispersed air flotation are the commonly used flotation techniques according to the method of bubble production. Dissolved air flotation is the most widely used method for the treatment of industrial effluent. Van Vuuren et al. (1965) performed a study on flotation and reported that flocculation requires several hours of sedimentation, while flotation shortens the duration to only a few minutes. The DAF procedure by chemical flocculation is reported to recover up to 6% (w/v) algal biomass slurries from algae culture (Bare et al., 1975). Although flotation has been used by several researchers as a potential harvesting method, there is only limited evidence of its technical and economic viability.

6.7 FLOCCULATION

Flocculation is used to separate microalgal cells from broth by the addition of one or more chemicals. Microalgal cell walls carry a negative charge that prevents self-aggregation within the suspension. This negative charge is countered by the addition of polyvalent ions called flocculants. These can be cationic, anionic, or nonionic, and they flocculate the cells without affecting their composition and/or being toxic. These flocculants have been classified into two groups, namely (1) inorganic agents, including polyvalent metal ions such as Al^{3+} and Fe^{+3} that form polyhydroxy complexes at suitable pH; and (2) polymeric flocculants, including ionic, nonionic, natural, and synthetic polymers. Among the former group, aluminum sulfate, ferric chloride, and ferrous sulfate are commonly used multivalent flocculants whose efficiency is directly proportional to the ionic charge. Fe^{3+} has been reported to be 80% efficient in harvesting different types of algae (Knuckey et al., 2006). The mechanism of polymer flocculation involves ionic interaction between polyelectrolyte and algal cells, resulting in the bridging of algae and formation of flocs. The extent of aggregation depends on the charge, molecular weight, and concentration of polymers. It has been observed that binding capability increases with an increase in molecular weight and charge on the polymer. Algal properties such as the pH of broth, concentration of biomass, and its charge are equally important to consider when selecting a polymer. Tenney et al. (1969) found effective flocculation in *Chlorella* when using a cationic polyelectrolyte, whereas an anionic polyelectrolyte failed to do so. Divakaran and Pillai (2002) successfully used chitosan as a bioflocculant for *Spirulina*, *Oscillatoria*, *Chlorella*, and *Synechocystis* spp. The efficiency of the method is affected by media pH, and best results were recorded at pH 7.0 for freshwater and a lower pH for marine species. Organic flocculants are reported to be beneficial in terms of their lower sensitivity to media pH, low dosage requirements, and wider range of applications. Heasman et al. (2000) also studied chitosan as a flocculant for *Tetraselmis chui*, *Thalassiosira pseudonana*, and *Isochrysis* sp., and they found that only 40 mg L^{-1} of chemical was needed for complete aggregation, whereas 150 mg L^{-1} was needed for *Chaetoceros muelleri*. Microbial flocculants (AM49) were also studied by Oh et al. (2001) for the harvesting of *Chlorella vulgaris*. This flocculant was found to be better than other commonly used flocculants. Recovery of more than 83% solids when operating in the pH range 5 to 11 was recorded; this is higher than that when using aluminum sulfate (72%) or the cationic polymer polyacrylamide (78%).

Algae also have the property of auto-flocculation when supplemental CO_2 supply is removed. Disruption of the CO_2 supply during photosynthesis increases the pH, which results in the precipitation of magnesium, calcium, phosphate, and carbonate salts along with algal cells. The positively charged ions interact with the negatively charged algal surfaces and bind them, resulting in auto-flocculation. Sukenik and Shelef (1984) conducted a study on auto-flocculation in pond and laboratory-scale experiments, and reported some very promising results. The unavailability of conducive conditions—especially light and CO_2—can, however, limit this process.

6.8   ELECTROLYTIC COAGULATION

The electrolytic coagulation (EC) process has recently been adapted by wastewater treatment plants for final polishing and removal of algae from partly treated wastewater. Active polyvalent metal anodes (usually iron or aluminum) are used to generate ionic flocculants such as Al^{3+} and Fe^{3+} ions. The latter agglomerate algae to form flocs due to the net negative charge and colloidal behavior of algal cells (Gao et al., 2010). The entire coagulation process involves the formation of coagulants by dissolution of the reactive anode, destabilization of colloidal suspensions, and aggregation of destabilized suspensions, resulting in the formation of algal flocs. The EC process is a more efficient chemical flocculation technique compared to conventional processes of direct interaction of the aluminum sulfate with algal suspensions (Aragón et al., 1992). The flocculated biomass is removed from water, either by sedimentation or flotation and skimming. For the latter to be effective, inactive metal cathodes are used to generate micro-gas (mainly hydrogen) bubbles that get entrapped in algae flocs and float them to the surface. Complete biomass removal from algal cultures having cell densities of 0.55×10^9 to 1.55×10^9 cells mL^{-1} has been reported using this process (Gao et al., 2010). However, the process may not be very effective for very dilute algal solutions because at low concentrations of total suspended solids (representing algal cells), the amount of colloids present in the culture solution may not be sufficient for significant amounts of settleable solids (Azarian et al., 2007).

6.9 ENERGY EFFICIENCIES OF HARVESTING PROCESSES

In terms of energy inputs, harvesting of algal biomass is the most energy-intensive process in biomass production. To date, there has been no specific commercial-scale algal harvesting technique that has been developed, and the approach has been to adapt separation technologies already in use in wastewater treatment and food processing industries. Therefore, the energy consumption and energy efficiency information available from those industries are discussed in this chapter to compare the energy efficiency of different algal harvesting techniques. The highest possible solids recovery (as %(w/v) total suspended solids (TSS)) and energy requirements for each of the harvesting processes are given in Table 6.1.

TABLE 6.1
Summary of Energy Usage and Highest Possible Solids (%w/v) Yields of Different Algae Harvesting Techniques

Harvesting Process	Highest Yield (% solids)	Energy Usage (kW-hm⁻³)	Ref.
Centrifugation	22.0	8.00	Girma et al., 2003
Gravity sedimentation	1.5	0.1	Shelef et al., 1984
Filtration (natural)	6.0	0.4	Semerjian et al., 2003
Filtration (pressurized)	27.0	0.88	Semerjian et al., 2003
Tangential flow filtration	8.9	2.06	Danquah et al., 2009
Vacuum filtration	18	5.9	Girma et al., 2003
Polymer flocculation	15.0	14.81	Danquah et al., 2009
Electro-coagulation	NA	1.5	Bektaş et al., 2004
Electro-flotation	5.0	5.0	Shelef et al., 1984; Azarian et al., 2007
Electro-flocculation	NA	0.331	Edzwald, 1995

Gravity sedimentation is relatively less energy intensive as fewer motors, pumps, and settling tanks are needed for its operation, thus resulting in low capital and operational cost and high expected life span. However, for a commercial-scale (>4 hectares) algae cultivation process and considering the slow sedimentation rates of algae, multiple tanks of large volumes (~100,000 L each) may be required.

Centrifugation is a very efficient technique, but the large energy requirement for the process clearly eliminates it as an option for harvesting a low-value energy crop. Therefore, this process can be ruled out for harvesting algae biomass for biofuel production, at least for first-stage dewatering (increasing solids content from algae culture on the order of 1%), on both cost and energy grounds.

Membrane filtration would be the next most efficient harvesting option; however, field experience on algae farms would be required to verify the lifetime and maintenance costs of the filter elements. Membrane filtration may be a competitive option if the back-flush function could be carried out with air knives in place of water jets, in order to achieve the desired consistency of 1% to 5% solids. Dissolved air flotation would be the expensive option in terms of cost and energy burden.

Of the flocculation-based processes, polymer flocculation is not only the most efficient, but also the most energy-intensive technique for dewatering. On the other hand, electro-flocculation techniques are cost-effective with low energy burden, but these techniques are still in their infancy and large-scale field testing is required to verify the overall process efficiency. Auto-flocculation is the lowest cost, lowest energy dewatering process by far, at one-tenth those of membrane filtration and polymer-based dewatering. Moreover, the chemicals required are pond nutrients, which can be recovered from the biomass for re-use either via anaerobic digestion, as would be the nitrogen, phosphorous, and potassium nutrients, or via an inexpensive carbonic acid extraction process if necessary, thereby avoiding the production-scale limitations imposed by synthetic polymer flocculants. As with the growth ponds and the anaerobic digesters, auto-flocculation employs managed natural processes to achieve its ends, at considerable savings in cost and energy.

Although the data presented in Table 6.1 appear straightforward and it would be easy for anyone to compare the harvesting and energy efficiencies of various processes, there are a variety of fundamental operational issues associated with each process. Therefore, it is important to carefully analyze several parameters, such as cell morphology, ionic strength of the media, pH, culture density, and final downstream processing of harvested biomass, when selecting a suitable harvesting technique. For example, very small sized algae could hinder the harvesting efficiency and would have a negative impact on the economics of biomass production if subjected to gravity sedimentation and filtration. However, if such algae could be made to float via the DAF (dissolved air flotation) flocculation process, this may facilitate harvesting. Furthermore, downstream processing of the harvested biomass to get final products will also be an important factor in selecting the harvesting process. If the biomass will be subjected to anaerobic digestion (AD) for biogas production, a solids content up to 5% (w/v) would suffice; whereas, if lipid extraction followed by biodiesel production is the goal, the biomass needs to be dewatered to lower moisture contents. There is considerable interest in efficient but less energy-intensive harvesting technologies to make microalgae cultivation cost effective and competitive for renewable bioenergy production. Thus far, no single harvesting technique can be universally applied to algae cultivation systems, and a combination of different techniques could be applied in a specific sequence to achieve maximum biomass concentration with minimum energy usage. Moreover, there could be considerable costs and energy savings in custom-designed, multi-stage harvesting techniques for algal farms, in which a variety of harvesting techniques are arranged in a specific sequence based on culture chemistry, and the specific characteristics of each technique and its energy requirements. Such systems can achieve dewatering of pond water to either 5%, or 10% to 20% solids at the least energy input and cost. In an open pond system, dominant algal species could range from small unicellular to large colonial or filamentous species. In such cases, TFF and other filtration techniques could be used as the first stage to remove filamentous and auto-flocculated algae, followed by chemical flocculation, sedimentation, and/ or flotation to produce algal slurries with 1% to 5% solids, which could be either directly subject to the AD process if biogas is the final biofuel or subject to centrifugation to achieve >20% solids. Centrifuging 5% algal slurry would reduce the energy and cost requirements for this technique by 100 times, as opposed to direct centrifugation of otherwise very dilute algae culture (~0.05% solids). Similarly, the pond water temperature, alkalinity, and pH may also vary during different climatic conditions throughout the year and even during different times of the day, thus impacting the ionic strength, salts solubility, and eventually the biomass auto-flocculation properties. Auto-flocculation could be the lowest-cost, lowest-energy dewatering process by far, at one tenth those of membrane filtration and polymer-based dewatering.

6.10 CONCLUSION

There are several biomass harvesting techniques available for the recovery of algae from culture broth. However, no individual technique can be applied ubiquitously due to technical and economical limitations. Gravity sedimentation is relatively less energy intensive but the slow sedimentation rates of algae may negatively impact production economics. Centrifugation is very efficient, but the large energy requirements for the centrifuge clearly eliminate it as an option for direct harvesting of a low-value energy crop. However, if used as a second-stage process for harvesting 5% algal slurries to higher solids concentrations, this could significantly reduce the energy and cost requirements. Membrane filtration is an efficient harvesting option; however, field experience on algae farms would be required to verify the lifetime and maintenance costs of filter elements. Dissolved air flotation would be an expensive option in terms of cost and energy. Polymer flocculation is also efficient but energy intensive for dewatering, while electro-flocculation is cost effective with low energy usage. Auto-flocculation is the lowest cost, lowest energy dewatering process by far. It is recommended to apply custom-designed multi-stage harvesting techniques for algal harvesting in which a variety of harvesting technologies are organized in some sequence to achieve the highest efficiency and lowest cost.

REFERENCES

Aragón, A.B., Padilla, R.B., and Ros de Ursinos, J.A.F. (1992). Experimental study of the recovery of algae cultured in effluents from the anaerobic biological treatment of urban wastewaters. *Resources, Conservation and Recycling*, 6: 293–302.

Azarian, G.H., Mesdaghinia, A.R., Vaezi, F., Nabizadeh, R., and Nematollahi, D. (2007). Algae removal by electro-coagulation process, application for treatment of the effluent from an industrial wastewater treatment plant. *Iranian Journal of Public Health*, 36: 57–64.

Bare, W.F., Jones, N.B., and Middlebrooks E.J. (1975). Algae removal using dissolved air flotation. *Journal of the Water Pollution Control Federation*, 47: 153–169.

Bektaş, N., Akbulut, H., Inan, H., and Dimoglo, A. (2004). Removal of phosphate from aqueous solutions by electro-coagulation. *Journal of Hazardous Materials*, 106: 101–105.

Benemann, J., and Oswald, W. (1996). Systems and Economic Analysis of Microalgae Ponds for Conversion of CO_2 to Biomass. Final Report to the U.S. Department of Energy. Pittsburgh Energy Technology Center, Grant No. DE-FG22-93PC93204.

Brennan, L., and Owende, P. (2010). Biofuels from microalgae—A review of technologies for production, processing, and extractions of biofuels and co-products. *Renewable and Sustainable Energy Reviews*, 14: 557–577.

Chisti, Y. (2007). Biodiesel from microalgae. *Biotechnology Advances*, 25: 294–306.

Danquah, M.K., Ang, L., Uduman, N., Moheimani, N., and Forde, G.M. (2009). Dewatering of microalgal culture for biodiesel production: Exploring polymer flocculation and tangential flow filtration. *Journal of Chemical Technology and Biotechnology*, 84: 1078–1083.

Divakaran, R., and Pillai, V.N.S. (2002). Flocculation of algae using chitosan. *Journal of Applied Phycology*, 14: 419–422.

Edzwald, J. (1995). Principles and applications of dissolved air flotation. *Water Science and Technology*, 31: 1–23.

Gao, S., Yang, J., Tian, J., Ma, F., Tu, G., and Du, M. (2010). Electro-coagulation–flotation process for algae removal. *Journal of Hazardous Materials*, 177: 336–343.

Girma, E., Belarbi E.H., Fernandez, G.A., Medina A.R., and Chisti Y. (2003). Recovery of microalgal biomass and metabolites: process options and economics. *Biotechnology Advances*, 20: 491–515.

Harun, R., Singh, M., Forde, G.M., and Danquah, M.K. (2010). Bioprocess engineering of microalgae to produce a variety of consumer products. *Renewable and Sustainable Energy Reviews*, 14: 1037–1047.

Heasman, M., Diemar, J., Connor, O'., Shushames, T., Foulkes, L., and Nell, J.A. (2000). Development of extended shelf-life microalgae concentrate diets harvested by centrifugation for bivalve molluscs—A summary. *Aquaculture Research*, 31: 637–659.

Khan, S.A., Hussain, M.Z., Prasad, S., and Banerjee, U.C. (2009). Prospects of biodiesel production from microalgae in India. *Renewable and Sustainable Energy Reviews*, 13: 2361–2372.

Knuckey, R.M., Brown, M.R., Robert, R., and Frampton, D.M.F. (2006). Production of microalgal concentrates by flocculation and their assessment as aquaculture feeds. *Aquacultural Engineering*, 35: 300–313.

Lee, S.J., Kim, S.B., Kim, J.E., Kwon, G.S., Yoon, B.D., and Oh, H.M. (1998). Effects of harvesting method and growth stage on the flocculation of the green alga *Botryococcus braunii*. *Letters in Applied Microbiology*, 27: 14–18.

Matis, K.A., Gallios, G.P., and Kydros, K.A. (1993). Separation of fines by flotation techniques. *Separations Technology*, 3: 76–90.

Mohn, F.H. (1980). Experiences and strategies in the recovery of biomass in mass culture of microalgae. In Shelef, G., and Soeder, C.J., Editors, *Algal Biomass: Production and Use*, Elsevier Science Ltd., Kidlington, UK, pp. 547–571.

Oh, H.M., Lee, S.J., Park, M.H., Kim, H.S., Kim, H.C., Yoon, J.H., Kwon, G.S., and Yoon, B.D. (2001). Harvesting of *Chlorella vulgaris* using a bioflocculant from *Paenibacillus* sp. AM49. *Biotechnology Letters*, 23: 1229–1234.

Pittman, J.K., Dean, A.P., and Osundeko, O. (2011). The potential of sustainable algal biofuel production using wastewater resources. *Bioresource Technology*, 102: 17–25.

Schenk, P. M., Thomas-Hall, S.R., Stephens, E., Marx, U.C., Mussgnug, J.H., Posten, C., Kruse, O., and Hankamer, B. (2008). Second generation biofuels: High efficiency microalgae for biodiesel production. *BioEnergy Research*, 1: 20–43.

Semerjian, L., and Ayoub, G.M. (2003). High-pH–magnesium coagulation–flocculation in wastewater treatment. *Advances in Environmental Research*, 7: 389–403.

Shelef, G., Sukenik, A., and Green, M. (1984). Microalgae Harvesting and Processing: A Literature Review. Report, Solar Energy Research Institute, Golden, CO, SERI Report No. 231-2396.

Sim, T.S., Goh, A., and Becker, E.W. (1988). Comparison of centrifugation, dissolved air flotation and drum filtration techniques for harvesting sewage-grown algae. *Biomass*, 16: 51–62.

Spolaore, P., Joannis-Cassan, C., Duran, E., and Isambert, A. (2006). Commercial applications of microalgae. *Journal of Bioscience and Bioengineering*, 101: 87–96.

Sukenik, A., and Shelef, G. (1984). Algal autoflocculation—Verification and proposed mechanism. *Biotechnology and Bioengineering*, 26: 142–147.

Taylor, G., Baird, D.J., and Soares, A.M.V.M. (1998). Surface binding of contaminants by algae: Consequences for lethal toxicity and feeding to *Daphnia magna* Straus. *Environmental Toxicology and Chemistry*, 17: 412–419.

Tenney, M.W., Echelberger, Jr., W.F., Schuessler, R.G., and Pavoni, J.L. (1969). Algal flocculation with synthetic organic polyelectrolytes. *Applied Microbiology*, 18: 965–971.

Uduman, N., Qi, Y., Danquah, M.K., Forde, G.M., and Hoadley, A. (2010). Dewatering of microalgal cultures: A major bottle to algal based fuels. *Journal of Renewable and Sustainable Energy*, 2: 012701–012715.

Van Vuuren, L.R.J., and Van Duuren, F.A. (1965). Removal of algae from wastewater maturation pond effluent. *Journal of the Water Pollution Control Federation*, 37: 1256–1262.

7 Lipid Identification and Extraction Techniques

Desikan Ramesh
Department of Farm Machinery
Agricultural Engineering College and Research Institute
Tamil Nadu Agricultural University
Coimbatore, Tamil Nadu, India

CONTENTS

7.1 LIPID QUANTIFICATION

Due to their high oil content, microalgae have attracted substantial research attention for biodiesel production and furthermore, algae have the capability to replace conventional biodiesel feedstocks. Algal strains collected from diverse aquatic environments require the evaluation of various important parameters such as oil content, lipid composition, growth rate, and metabolic efficiency under different conditions. One can decide whether the selected algal strain is suitable/unsuitable for biodiesel production based on the preliminary lipid analysis (both lipid yield and lipid composition).

Microalgal strains have the potential to produce up to 50% lipid by dry cell weight, depending on the species and specific growth conditions (Chisti, 2007). The neutral lipids present in microalgae are primarily in the form of triacylglycerols (TAGs). TAGs can be converted to fatty acid methyl esters (FAMEs) via transesterification. Recovery of the accumulated algae lipids from algae paste is generally carried out after rupturing the cells to free the lipids. Different cell disruption techniques are used to rupture the algae cells, including autoclaving, microwave, sonication, osmotic shock, and bead beating. Lee et al. (2010) evaluated five different cell disruption techniques for enhancing lipid extraction efficiency. They reported that the microwave oven method is an efficient method for extracting lipids from microalgae. Because of its simplicity and cost effectiveness, solvent extraction is widely used by researchers (Letellier and Budzinski, 1999). For laboratory-scale studies, lipid content and composition can be determined using well-established techniques. The most commonly used method for lipid extraction is the Bligh and Dyer method, or some variation thereof. Methods for simultaneous extraction and transesterification of algal biomass to extract the algal lipids are also available (Belarbi et al., 2000; Lewis et al., 2000).

7.2 LIPID PROFILES

7.2.1 IDENTIFICATION OF ALGAE LIPID PROFILES

The steps involved in the upstream process for algal oil-based biodiesel production include strain identification, optimization for higher lipid yield, and mass production. For quantifying the lipids available in microalgae, any one of the following chromatography techniques may be followed: high-pressure liquid chromatography (HPLC), gas chromatography (GC), or chromatography–mass spectrometry (LC/GC-MS). Usually, GC is employed for analyzing the algae lipid profiles after conversion to FAMEs. GC-FID (flame ionization detection) or GC-MS may be used for the identification of fatty acid profiles of algae lipids. In case of GC-FID, the retention time for the individual components of FAMEs is compared to known standards (Mansour, 2005; Lin et al., 2007; Paik et al., 2009). The lipid profile of algal oils can be analyzed via GC-FID using the ASTM D6584 and EN 14105 standard methods. The methylated lipid/FAMEs of algae may contain traces of contaminants such as chlorophylls, catalyst, or water, and samples injected in GC must be free from these contaminants to prevent GC column damage.

7.2.2 SUITABILITY OF ALGAE LIPID FOR BIODIESEL PRODUCTION

In general, algal oils contain a higher degree of polyunsaturated fatty acids (PUFA) (i.e., more than four double bonds) than vegetable oil (Belarbi et al., 2000; Harwood and Guschina, 2009) and higher free fatty acid content (>2%). ASTM D6751 (United States) and EN 14214 (European Union) provide the specifications for pure biodiesel (B_{100}, Table 7.1) and are used in many parts of world for comparing the fuel properties of biodiesel. The biodiesel standards developed in many countries are based on the availability of region-specific biodiesel feedstocks. The specifications developed for ensuring biodiesel fuel quality are frequently subjected to modifications, and biodiesel-producing countries are required to update their specifications according to changes in ASTM- or EN-based biodiesel standards.

The lipid composition of algal oil is different from plant oils/animal fats, and it varies with species and growing conditions (Mutanda et al., 2011). Major fuel and chemical properties considered for the selection of an alternate diesel fuel are

TABLE 7.1

Biodiesel (B$_{100}$) Standards Used in the United States (ASTM D6751) and Europe Union (EN 14214)

	Biodiesel Standards			
Parameters	EN Method	Values	ASTM Method	Values
Density, 15°C	EN 3675, EN 12185	860–900 kg m^{-3}	–	–
Kinematic viscosity, 40°C	EN 3104/3105	3.5–5.0 mm^2 s^{-1}	ASTM D445	1.9–6.0 mm^2 s^{-1}
Flash point	EN 3679	120°C (min.)	ASTM D93	93°C (min.)
Sulfur content	EN 20846	10 ppm	ASTM D5453	15 ppm
Carbon residue	EN 10370	0.30% (mol mol^{-1}) (max.)	ASTM D4530	0.050% mass (max.)
Cetane number	EN 5165	51 (min.)	ASTM D613	47 (min.)
Copper strip corrosion	EN 2160	No. 1	ASTM D130	No. 3 (max.)
Phosphorous content	EN 14107	10.0 mg kg^{-1} (max.)	ASTM D4951	0.001% mass (max.)
Oxidation stability	EN 14112	6.0 h (min.)	EN 14112	3.0 h (min.)
Acid value	EN 14104	0.50 mg KOH g^{-1} (max.)	ASTM D664	0.50 mg KOH g^{-1} (max.)
Free glycerol	EN 14105, EN 14106	0.020% (mol mol^{-1}) (max.)	ASTM D6584	0.020% mass
Total glycerol	EN 14105	0.25% (mol mol^{-1}) (max.)	ASTM D6584	0.24% mass

kinematic viscosity (KV), higher heating value (HHV), Cetane Number (CN), density, flashpoint, cold flow properties (cloud and pour points), carbon residue, oxidation stability, ash content, ignition quality, acid value (AV), saponification value (SV), and iodine value (IV). These properties can be compared with well-established international fuel standards for ensuring fuel quality for diesel engine applications. Properties such as SV, IV, and CN are considered more important for assessing alternate diesel fuels because they give basic information about the ignition quality of fuel, the presence of unsaturated fatty acids (UFAs), and the ignition properties of FAMEs, respectively. The higher iodine values of algal oil indicate the presence of higher UFAs, and heating these UFAs may be lead to the formation of deposits due to the polymerization of glycerides at high temperatures (Mittelbach, 1996; Ramos et al., 2009). The algal oils/FAMEs containing higher degrees of unsaturation are not recommended for biodiesel. The values of the SV, IV, and CN can be easily calculated from the lipid profiles of algal oil using equations developed by Krisnangkura (1986) and Kalayasiri et al. (1996).

Based on the lipid profiles of identified strains/species grown on a laboratory scale, they can be easily screened for suitability in biofuel production. Based on correlations developed between fatty acid compositions and the fuel properties of oils, the fuel quality of biodiesel derived from selected algal oils can be predicted through the lipid composition. Hence, it is necessary to determine the fatty acid profiles of extracted algal lipids for suitability in biodiesel production and assess the fuel quality. It is essential for biodiesel derived from microalgal oil to meet ASTM (2008) or EN (2003) biodiesel standards for ensuring fuel quality.

7.3   OIL EXTRACTION

The mass production of microalgae can be achieved through raceway ponds or photobioreactors. The feasibility of biodiesel production from algae is totally dependent on the technologies used in the downstream processing of algae. Downstream processing of algae involves dewatering, drying, oil extraction, biofuel production, and by-product utilization. Dewatering of algae is an energy-consuming process that requires a high capital investment for equipment. There is a greater possibility of quick spoilage of the harvested biomass slurry (5% to 15% dry solids) under hot climatic conditions due to its high moisture content (Brennan and Owende, 2010). The harvested slurry must be dewatered and dried quickly after harvesting. Drying is an important process in the downstream process that enhances biomass shelf life and lipid recovery efficiency. The process employed for oil extraction from dried algae is similar to that of oil extraction from oil seeds. In order to achieve economically viable algal oil-based biodiesel production, the technical feasibility must be thoroughly studied; that is, downstream technologies/processes must be optimized.

The oil content of dried algal biomass varies from 20% to 50% oil by weight and can be increased by optimizing growth parameters (Hu et al., 2008). Before commencing oil extraction from microalgae, it must be dewatered and dried to remove the moisture. The present oil extraction techniques applicable for wet/dry algae biomass have limitations due to technical barriers, difficulty in scaling-up, high cost investment, and extraction efficiency (Cooney et al., 2009). The choice of oil extraction technology depends on the moisture content, quantity to be treated, quality of the end-product, extraction efficiency, safety aspects, and cost economics. The main outputs from the oil extraction process are oil and oil cake. The extracted oil can be used for biodiesel production with/without pretreatment, depending on the quality of the oil. The actions involved in the oil extraction process are (1) breaking the algae cell walls, (2) freeing the oil, and (3) separating out the oil and oil cake. The most common methods used for the oil extracted from algae biomass are mechanical press, solvent extraction, and supercritical fluid extraction. The methods employed for oil extraction from microalgae are depicted in Figure 7.1.

7.3.1   MECHANICAL EXTRACTION

In the case of mechanical extraction, the feedstock (oil seed or algae biomass) is subjected to high pressure for rupture and release of the oil. The added advantages of mechanical extractions are that (1) no chemicals are used for extraction, (2) the process is free of chemicals in the products, and (3) the product is safe for storage. The major drawbacks of mechanical extraction are an inadequacy in complete lipid recovery from the feedstock, and the high energy inputs.

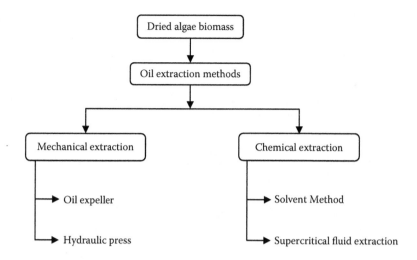

FIGURE 7.1 Classification of oil extraction methods used for dried algae biomass.

7.3.1.1 Oil Expeller

The first successful, continuously operating mechanical expeller was invented and patented by Anderson in the year 1900. An oil expeller is also called a screw press. The screw press is well suited for feedstocks having more than 30% oil content. The working principle of this machine is introducing pressure for crushing and breaking the cells, followed by squeezing out the algal oil. An oil press is the simplest method used for oil extraction from algae biomass (Popoola and Yangomodou, 2006; Demirbas, 2009). The oil cake obtained from a screw press may contain some amount of residual oil (4% to 5% by weight). The oil extraction efficiency improves with applied pressure in a particular range, but too much pressure leads to less lipid recovery, more heat generation, and choking problems. Increased heat produced due to excessive pressure applied inside the screw press leads to darkening of the oil and a low-quality oil cake. The major drawbacks of this method are the high energy consumption, high maintenance cost with low capacities, labor intensiveness, long extraction time and less efficiency than other methods.

7.3.2 Chemical Extraction

7.3.2.1 Solvent Extraction

The process of extracting oil from oil-containing materials using a suitable solvent is called *solvent extraction*. It is well suited for lipid recovery from materials with low oil content, and it produces oil cake with low residual oil content (<1% by weight) (Erickson et al., 1984; Hamm and Hamilton, 2000). Algae cell walls are made of multiple layers and they are more recalcitrant than those of other microorganisms (Sander and Murthy, 2009). Some species having an additional trilaminar sheath (TLS) containing an algaenan component are resistant to degradation (Allard et al., 2002; Versteegh and Blokker, 2004).

The solvent extraction process occurs when a solvent comes into contact with microalgae to release lipids, solvate the lipids in solvent, and separate the oil from miscella by distillation of the solvent. The drawbacks of solvent-based oil extraction are that (1) the solvent is highly inflammable, (2) the energy requirements are high, and (3) the process requires high capital investments.

A solvent extraction plant consists of extractors, desolventizers, evaporators, stripping towers, and condensers in the extractor. Solvent is sprayed over the oil-bearing materials and solvent penetrates the biomass, targeting the soluble compounds. The mixture of algae lipid dissolved in the solvent is called a miscella and is sent to collection tanks. The algal oil is separated from the solvent using evaporators and stripping towers. The oil coming out of these units is first cooled, and then filtered and sent to storage tanks. The oil cake coming out of the extractor unit may contain some residual amounts of solvent. The residual solvent present in the oil cake is removed in the desolventizer unit, and recovered solvent can be reused in the extractor.

The various organic chemical solvents employed for oil extraction include benzene, hexane, cyclohexane, acetone, chloroform, ethanol (96%), and hexane–ethanol (96%) combinations. It is possible to extract up to 98% quantitative of purified fatty acids (Richmond, 2004). Among these solvents, hexane is most commonly used in the food industry. Hexane meets many of the requirements of an ideal oil solvent (Johnson and Lusas, 1983) due to it having a high extraction efficiency, a low viscosity, and a low boiling point, and being a nonpolar solvent, is easily miscible with oil and inexpensive.

7.3.2.2 Supercritical Fluid Extraction (SFE)

When a fluid is subjected to temperatures and pressures above its critical temperature and pressure, it becomes a supercritical fluid. Supercritical fluid extraction (SFE) is the process of extracting oil from oil-containing materials using a supercritical fluid as the extraction solvent. The advantage of supercritical fluids used in oil extraction is their increased solvating power (Mercer and Armenta, 2011). Factors to consider when selecting an SFE solvent include that the solvent is nonflammable, nontoxic, has low critical parameters, good solvating properties, is easily separated from product, and is environmentally friendly and inexpensive. The added advantages of SFE over conventional solvent extraction are that it provides simple and flexible process control of temperature, shorter extraction times, low cost, and solvent-free product. The SFE consists of an SFE solvent tank, solvent and feed pumps, a high-pressure pump, extraction vessels, and restrictor and absorbent vessels. Carbon dioxide (CO_2) is widely used as a solvent in SFE due to its moderate critical temperature (31.1°C) and pressure (72.9 atm) (Cooney et al., 2009). In this method, CO_2 is used as the extracting solvent when it is in a supercritical state (i.e., it has both gas and liquid properties). The supercritical state of CO_2 can be achieved by liquefying CO_2 under higher pressure and heating to a particular temperature. The important operating parameters considered for optimizing the extraction efficiency of this method are operating temperature and pressure, quantity of CO_2 supplied, feed particle size, and residence time. Dried algae paste must be used for supercritical extraction; this helps in increasing the contact time between the SFE solvent and the algae paste. CO_2 acts as a gas in air at ambient temperature, and can be removed after the

TABLE 7.2

Comparison of different oil extraction methods

S. No.	Parameters	Mechanical Pressing	Solvent Extraction	Supercritical Fluid Extraction	Ref.
1	Algae form	Dry	Dry	Dry	Sahena, 2009; Singh and Sai, 2011; de Boer et al., 2012
2	Pretreatment to algae biomass for cell rupture	Required to improve lipid extraction efficiency	Required to improve lipid extraction efficiency	Not required	Cooney et al., 2009; Sahena et al., 2009; Mercer and Armenta, 2011; de Boer et al., 2012
3	Solvent used	No solvent	Hexane	Supercritical CO_2	Popoola and Yangomodou, 2006; Cooney et al., 2009; Demirbas, 2009
4	Working fluid	Not applicable	Toxic	Nontoxic	Mercer and Armenta, 2011
5	Oil recovery (%)	70–75	96	100	Demirbas, 2009; Popoola and Yangomodou, 2006; Demirbas and Demirbas, 2010
6	Purity of oil	NA	Both polar and nonpolar and colors	Only nonpolar and colors	Cooney et al., 2009; Mercer and Armenta, 2011
7	Energy consumption	Low	High	High	Macias-Sanchez et al., 2005; Cooney et al., 2009; Mercer and Armenta, 2011; de Boer et al, 2012
	Capital investment	Low	High	High	Mercer and Armenta, 2011
8	Recycling unit	Nil	Necessary for reuse	Necessary for reuse	—
9	Scaling up	Possible	Possible	Difficult at this time	Pawliszyn, 1993; Macias-Sanchez et al., 2005; Mercer and Armenta, 2011

extraction and reused again for further extractions. A comparison of three different oil extraction methods used for lipid recovery is provided in Table 7.2.

7.4 CONCLUSION

Microalgae are becoming more attractive feedstocks for biodiesel production as higher oil-yielding algae have the potential to replace conventional biodiesel feedstocks. The viability of microalgae oil-based biodiesel production primarily depends on the identification of appropriate higher lipid producing algal strains. From preliminary studies on lipid analysis for identifying algae, important fuel properties of the algal oil can be predicted and compared with biodiesel standards. The simplest method of assessing the fuel quality of biodiesel is predicting its fuel properties based on the fatty acid composition of algal oil, thereby allowing us to ascertain the suitability of selected algae strains for biodiesel production. The economical/technical viability of microalgal oil-based biodiesel production depends on implementation of the suitable technologies used in the downstream processing of microalgae.

REFERENCES

Allard, B., Rager M.N., and Templier, J. (2002). Occurrence of high molecular weight lipids (c80+) in the trilaminar outer cell walls of some freshwater microalgae. A reappraisal of algaenan structure. *Organic Geochemistry*, 33: 789–801.

Anderson, V.D. (1900). Press, U.S. Patent 647, 354.

ASTM Standard Specification for Biodiesel Fuel (B100) Blend Stock for Distillate Fuels, (2008). In *Annual Book of ASTM Standards*, ASTM International, West Conshohocken, PA, Method D6751-08.

Belarbi, E. H., Molina, E., and Chisti, Y. (2000). A process for high yield and scalable recovery of high purity eicosapentaenoic acid esters from microalgae and fish oil. *Enzyme Microbiology and Technology,* 26: 516–529.

Brennan, L., and Owende, P. (2010). Biofuels from microalgae. A review of technologies for production, processing, and extractions of biofuels and co-products. *Renewable and Sustainable Energy Review,* 14: 557–577.

Chisti, Y. (2007). Biodiesel from microalgae. *Biotechnology Advances*, 25: 294–306.

Cooney, M., Young, G., and Nagle, N. (2009). Extraction of bio-oils from microalgae. *Separation and Purification Reviews,* 38: 291–325.

De Boer, K., Moheimani, N.R., Borowitzka, M.A., and Bahri, P.A. (2012) Extraction and conversion pathways for microalgae to biodiesel: A review focused on energy consumption. *Journal of Applied Phycology,* 24: 1681–1698.

Demirbas, A. (2009). Production of biodiesel from algae oils. *Energy Sources, Part A,* 31: 163–168.

Demirbas, A., and Demirbas, M.F. (2010). *Algae Energy: Algae as a New Source of Biodiesel.* Springer London Ltd., United Kingdom, p. 143.

EN Committee for Standardization Automotive Fuels (2003). Fatty Acid Methyl Esters (FAME) for Diesel Engines—Requirements and Test Methods. European Committee for Standardization, Brussels. Method EN 14214.

Erickson, R.D., Pryde, H.E., Brekke, L.O., Mounts, L.T., and Falb, A.R. (1984). *Handbook of Soy Oil Processing and Utilization.* American Oil Chemists Society, Champaign, IL, p. 598.

Hamm, W., and Hamilton, J.R. (2000). *Edible Oil Processing.* Boca Raton, FL: CRC Press LLC, p. 281.

Harwood, J.L., and Guschina, I.A. (2009). The versatility of algae and their lipid metabolism. *Biochimie,* 91: 679–684.

Hu, Q., Sommerfeld, M., Jarvis, E., Ghirardi, M.L., Posewitz, M.C., and Seibert, M. (2008). Microalgal triacylglycerols as feedstocks for biofuel production. *The Plant Journal*, 54: 621–639.

Johnson, L.A., and Lusas, E.W. (1983). Comparison of alternative solvents for oils extraction. *Journal of the American Oil Chemists Society,* 60: 229–242.

Kalayasiri, P., Jayashke, N., and Krisnangkura, K. (1996). Survey of seed oils for use as diesel fuels. *Journal of the American Oil Chemists Society,* 73: 471–474.

Krisnangkura, K. (1986). A simple method for estimation of cetane index of vegetable oil methyl esters. *Journal of the American Oil Chemists Society,* 63: 552–553.

Lee, J., Yoo, C., Jun, S., Ahn, C., and Hee-Mock, O. (2010). Comparison of several methods for effective lipid extraction from microalgae. *Bioresource Technology,* 101(Suppl. 1): S75–S77.

Letellier, M., and Budzinski, H. (1999). Microwave assisted extraction of organic compounds. *Analusis,* 27: 259.

Lewis, T., Nichols, P.D., and McMeekin, T.A. (2000). Evaluation of extraction methods for recovery of fatty acids from lipid producing micro-heterotrophs. *Journal of Microbiological Methods,* 43: 107–116.

Lin, Y.H., Chang, F.L., Tsao, C.Y., and Leu, J.Y. (2007). Influence of growth phase and nutrient source on fatty acid composition of *Isochrysis galbana* CCMP 1324 in a batch photoreactor. *Biochemical Engineering Journal,* 37: 166–176.

Macias-Sanchez, M.D., Mantell, C., Rodríguez, M., Martínez de la Ossa, E., Lubián, L.M., and Montero, O. (2005). Supercritical fluid extraction of carotenoids and chlorophyll a from *Nannochloropsis gaditana. Journal of Food Engineering,* 66: 245–251.

Mansour, M.P. (2005). Reversed-phase high-performance liquid chromatography purification of methyl esters of C16-C28 polyunsaturated fatty acids in microalgae, including octacosaoctaenoic acid [28:8(n–3)]. *Journal of Chromatography A,* 1097: 54–58.

Mercer, P., and Armenta, R.E. (2011). Developments in oil extraction from microalgae. *European, Journal of Lipid Science and Technology,* 113: 539–547.

Mittelbach, M. (1996). Diesel fuel derived from vegetable oils. VI. Specifications and quality control of biodiesel. *Bioresource Technology,* 56: 7–11.

Mutanda, T., Ramesh, D., Karthikeyan, S. Kumari, S., Anandraj A., and Bux, F. (2011). Bioprospecting for hyper-lipid producing microalgal strains for sustainable biofuel production. *Bioresource Technology,* 102: 57–70.

Paik, M.J., Kim, H., Lee, J., Brand, J., and Kim, K.R. (2009). Separation of triacylglycerols and free fatty acids in microalgal lipids by solid phase extraction for separate fatty acid profiling analysis by gas chromatography. *Journal of Chromatography A,* 1216: 5917–5923.

Pawliszyn, J. (1993). Kinetic model of supercritical fluid extraction. *Journal of Chromatography Science,* 31: 31–37.

Popoola, T.O.S., and Yangomodou, O.D. (2006). Extraction, properties and utilization potentials of cassava seed oil. *Biotechnology,* 5: 38–41.

Ramos, M.J., Fernandez, C.M., Casas, A., Rodríguez, L., and Perez, A. (2009). Influence of fatty acid composition of raw materials on biodiesel properties. *Bioresource Technology,* 100: 261–268.

Richmond A. (2004). *Handbook of Microalgal Culture: Biotechnology and Applied Phycology.* Blackwell Science Ltd., Malden, MA.

Sahena, F., Zaidul, I.S.M., Jinap, S., Karim, A.A., Abbas, K.A., Norulaini, N.A.N., and Omar, A.K.M. (2009). Application of supercritical CO_2 in lipid extraction – A review. *Journal of Food Engineering,* 95: 240–253.

Sander, K., and Murthy G.S. (2009). Enzymatic degradation of microalgal cell walls. *ASABE Annual International Meeting,* Sponsored by ASABE Grand Sierra Resort and Casino, Reno, Nevada, June 21–June 24, 2009. Paper number 1035636.

Singh, J., and Gu, S. (2010). Commercialization potential of microalgae for biofuels production. *Renewable and Sustainable Energy Reviews,* 14: 2596–2610.

Versteegh, G.J.M., and Blokker, P. (2004). Resistant macromolecules of extant and fossil microalgae. *Phycology Research,* 52: 325–339.

8 Synthesis of Biodiesel/ Bio-Oil from Microalgae

Bhaskar Singh
Department of Applied Chemistry
Indian Institute of Technology (BHU)
Varanasi, India

Yun Liu
College of Life Science and Technology
Beijing University of Chemical Technology
Beijing, China

Yogesh C. Sharma
Department of Applied Chemistry
Indian Institute of Technology (BHU)
Varanasi, India

CONTENTS

8.1 INTRODUCTION

Microalgae have emerged as a potential feedstock for the production of biodiesel. The steps involved in the production of bio-oil and biodiesel from microalgae include cultivation, harvesting, dewatering and concentrating microalgae, extraction of oil/lipids from the microalgae, and separation of triglycerides and free fatty acids from the crude lipids (for the synthesis of biodiesel). The final step consists of pyrolysis or thermochemical catalytic liquefaction for the production of bio-oil and esterification or transesterification for the synthesis of biodiesel. The species of microalgae employed for the production of biofuel include *Chlorella vulgaris*, *Chlorella sorokiniana*, *Sargassum patens* C., and *Spirulina*. The method of biodiesel production from algal biomass can be done either by direct transesterification or in two steps involving the extraction of oil from algae followed by transesterification. Economic *in situ* transesterification of the microalgae has been adopted that involves combining the two steps of lipid extraction and transesterification into a single step. Direct transesterification of the microalgae after cell disruption by sonication resulted in a high conversion of biodiesel (97.25%). The fuel properties of the biodiesel synthesized from the microalgal oil derived from *Chlorella protothecoides* showed high fuel quality with a cold filter plugging point of −13°C. A high composition of unsaturated fatty acid methyl ester content in the microalgal oil methyl esters (MOME) (i.e., 90.7 wt%) led to a low oxidation stability of the fuel (4.5 h). The chemical treatment, pyrolysis, or thermochemical catalytic liquefaction of microalgal oil for the synthesis of bio-oil eliminates the dewatering and drying steps. The major constituents of bio-oil obtained from brown microalgae *Sargassum patens* C. Agardh by hydrothermal liquefaction consist of carbon (64.64%), followed by oxygen (22.04%), hydrogen (7.35%), nitrogen (2.45%), and sulfur (0.67%).

The alga belongs to the third-generation feedstock for the synthesis of a renewable fuel, biodiesel, or bio-oil. The first generation of feedstock was the crop species, and second-generation feedstock consisted of grasses and trees, which principally consisted of lignocellulosic biomass. With the limited availability of crop species, the focus of recent research has been on second- and third-generation feedstocks (Stephenson et al., 2011). Second-generation feedstocks have certain constraints that involve breaking the complex structure of lignin and converting the crystalline cellulose to amorphous cellulose. The process involved in second-generation biofuels makes it quite energy intensive. Hence, the focus of the research to a large extent in recent years has been on third-generation feedstocks, that is, microalgae (Lam and Lee, 2012). The conversion of microalgal lipids into biodiesel is a holistic approach that begins with the identification of an appropriate microalgal species that has a high potential to accumulate oil within the cells. The oil consists of crude lipids and neutral lipids. Neutral lipids consist of triglycerides, free fatty acids, hydrocarbons, sterols, wax and sterol esters, and free alcohols. Among these, only triglycerides and free fatty acids are saponifiable, and hence can be converted to biodiesel by esterification or transesterification. Crude lipids consist of neutral lipids along with pigments (Sharma et al., 2011). The triglycerides and free fatty acids are the part of microalgal lipids that can be converted to biodiesel or bio-oil. The microalgal biomass can be used for the production of biofuel, either by pyrolysis or through direct combustion or thermochemical liquefaction in which bio-oil is produced. Alternatively, the lipid can be derived from microalgal biomass and converted to biodiesel via transesterification (Kao et al., 2012).

In general, microorganisms that accumulate more than 20% to 25% of their weight as lipid are called *oleaginous species* (Kang et al., 2011). As these oleaginous microorganisms can accumulate a large amount of oil within their cells, they can be very good feedstocks from which to extract oil and lipids that can be converted to biofuel (bio-oil or biodiesel). In some of the algae, the lipid content may be as high as 75% of their dry biomass. Most species of algae produce triglycerides (that can be utilized for biofuel production) and alkanes. A few algal species may also contain long-chain hydrocarbons that are formed via the terpenoid pathway (Sivakumar et al., 2012). Hence, the characterization of the algae is prerequisite to assessing the potential of the microalgae extract to be converted to biofuel. Once the triglyceride content in the microalgal species is determined, depending on their feasibility, the microalgae extract can be utilized for its conversion to either bio-oil or biodiesel. This chapter deals with the conversion of microalgal lipids into biodiesel or bio-oil, taking into account the following aspects: esterification/transesterification, and chemical method.

8.2 ESTERIFICATION/TRANSESTERIFICATION

Batan et al. (2010) reported that the net energy ratio of biodiesel produced from microalgae (*Nannochloropsis* species) was found to be 0.93 MJ (MJ = megajoule) of energy needed to produce 1 MJ of energy. The major advantage of using microalgae-derived biodiesel is the reduction in CO_2 equivalent emissions amounting to 75 g MJ^{-1} of energy produced. Liu et al. (2012) also report that biodiesel from algae has a positive energy impact and estimate 1.4 MJ of energy production per megajoule consumption of energy. In addition, there will be reduction of 0.19 kg CO_2-equivalents per kilometer of travel by transport.

Johnson and Wen (2009) adopted two methods for the production of biodiesel from a heterotrophic microalga, *Schizochytrium limacinum*. The first method adopted was direct transesterification of algal biomass using wet and dry biomass separately. The second method consisted of two steps involving extraction of oil from algae, followed by transesterification of wet and dry biomass. When the direct transesterification method was used, the yield of biodiesel obtained was greater than 66% for wet as well as dry biomass. However, the fatty acid methyl ester (FAME) content in the wet biomass was found to be very low (7.76%) in comparison with that from dry biomass (63.47%). A 57% crude biodiesel yield was obtained through the two-step method with a FAME content of 66.37%. Using the wet biomass, the FAME content of biodiesel was 52.66%. The one-stage direct transesterification method used various solvents (e.g., chloroform, hexane, and petroleum ether) to treat the algal biomass. However, a comparatively higher content of FAME was observed when only chloroform was used as the solvent. It has been found that direct transesterification is preferable instead of the conventional steps involved in the production of biodiesel from microalgae (i.e., extraction of oil from microalgae and transesterification of the expelled oil) as the production cost of the fuel will be reduced. However, the drying of algal biomass was found to be a prerequisite in order to obtain a high yield and conversion of biodiesel when using direct transesterification of the microalgae. To prevent oxidation of the unsaturated FAME in biodiesel, Johnson and Wen (2009) added 100 ppm butylated hydroxytoluene to the biodiesel. However, the fuel did not meet the European standards (EN 14103) specifications, which specify that the ester content in biodiesel must be at least 96.5% (Sarin et al., 2009).

Vijayaraghavan and Hemanathan (2009) reported on the production of biodiesel from freshwater algae. A high lipid content of 45 ± 4% was obtained from the microalgae and was used for the synthesis of biodiesel by transesterification using methanol as a reactant and KOH as a catalyst. Upon transesterification, the fuel properties of biodiesel were determined. The acid value of the biodiesel (0.40 mg KOH g^{-1}) was within ASTM (American Society for Testing & Materials) D 6751 specifications for biodiesel. The values of the other important parameters (i.e., density, ash, flash point, pour point, calorific value, cetane number, water content, and copper strip corrosion) were characterized. It was found that some of the parameters (i.e., density, ash, flash point, and water content) did not meet the ASTM specifications. The density of biodiesel was found to be low (801 kg m^{-3}), whereas the Indian specifications have a range of 860 to 900 kg m^{-3}. The ash content of the biodiesel was 0.21 mass%, in contrast to the 0.01% specified by Indian standards. Similarly, the flash point also had a low value of 98°C instead of the minimum value of 120°C. The water content (<0.02 vol.%) was also slightly higher than specifications (<0.03 vol.%). However, other parameters (i.e., pour point, calorific value, and cetane number) were found to be within Indian specifications.

An in-situ transesterification method has been adopted by Velasquez-Orta et al. (2012) for the transesterification of microalgae, *Chlorella vulgaris*. The *in situ* transesterification of this microalga was performed by combining the two steps of lipid extraction and transesterification into a single step. Although the reaction ran to completion in less time (75 min) using NaOH as the catalyst, a low conversion of FAME (77.6 ± 2.3 wt%) was obtained that does not meet the specifications of the European Union (EN 14103), which specifies that the ester content must be at least 96.5% (Sarin et al., 2009). However, when an acid catalyst (sulfuric acid) was used, a high FAME yield of 96.8 ± 6.3 wt% was obtained although a longer reaction time (20 h) was required. Also, a high methanol ratio (600:1) was employed, which will escalate the production cost of biodiesel. Tran et al. (2012) produced biodiesel from *Chlorella vulgaris* (ESP-31) using an enzyme (*Burkholderia* lipase) as a heterogeneous catalyst. The biodiesel was synthesized in two ways: (1) transesterification of microalgal oil, and (2) direct transesterification of the microalgae after disruption of its cells by sonication. A moderate conversion (72.1%) of the microalgae to biodiesel was obtained with the first method, whereas a high conversion (97.25%) was obtained using the second method. The immobilized enzyme was reused for six runs without any significant loss in catalytic activity. Being catalyzed by an enzyme, the catalyst was found to function even in the presence of water (>71.39 wt%). However, a higher molar ratio (67.93, methanol to oil) was needed to achieve an ester conversion of greater than 96 wt% of oil that will escalate the production cost of the biodiesel.

The fuel properties of the biodiesel synthesized from the microalgal oil derived from *Chlorella protothecoides* has been investigated by Chen et al. (2012). The microalgal oil methyl ester (MOME) with a high ester content (97.7%) demonstrated development of biodiesel of high fuel quality with a cold filter plugging point of −13°C, which is an indication that the fuel may be used even under extremely cold conditions. The viscosity of the MOME was 4.43 mm^2 s^{-1} at 40°C, which is within the specifications for biodiesel specified by the ASTM. However, the oxidation stability of the fuel was low (4.5 h), which was due to high amounts of unsaturated fatty acid content in the MOME (i.e., 90.7 wt%). The induction time as per the Indian and European specifications is at least 6 h (Sarin et al., 2009). Hence, it had been recommended that the MOME should use up to 20 vol% blended with mineral diesel. A higher blend of biodiesel in mineral diesel will require the addition of antioxidants so that the fuel does not get oxidized rapidly and remains within specifications. Siegler et al. (2012) extracted oil from the microalgae *Auxenochlorella protothecoides* and found it to be a potential source for biodiesel production. The degree of unsaturation (DU) in the microalgal oil, which is a measure of the unsaturated fatty acid content, was determined to be 137. Using the DU, the cold filter plugging point value of the biodiesel was expected to be −12°C, which can support the use of fuel even in cold climatic conditions.

Lardon et al. (2009) observed that despite biodiesel derived from microalgae having immense potential to provide an alternative source of fuel, energy and fertilizer consumption should be reduced for its economic viability. Using *Chlorella vulgaris* as a model species, it has been found that a substantial portion of energy consumption amounting to 70% and 90% of the total energy is used for lipid extraction when using wet and dry biomass, respectively. Hence, technologies must be developed for economical extraction of oil from microalgal cells. Rosch et al. (2012) advocate the reuse of residual algal biomass after oil extraction for the supply of nutrients, which according to estimates may vary from 0.23 to 1.55 kg nitrogen and 29 to 145 g phosphorous (depending on the cultivation conditions of microalgae) for the production of 1 L biodiesel.

8.3   THERMOCHEMICAL

The major cost attributed to the production of biodiesel is the dewatering and drying step, which consumes 9 to 16 GJ of energy per ton of biodiesel produced (Chowdhury et al., 2012). The dewatering and drying step can be negated if the microalgal oil is subjected to pyrolysis for the synthesis of bio-oil as a biofuel. The thermochemical method adopted for the preparation of fuel from microalgae is through pyrolysis, where the organic compound is thermally decomposed at a high temperature in the absence of oxygen. Zou et al. (2009) produced bio-oil by thermochemical catalytic liquefaction of *Dunaliella tertiolecta*. A yield of 97.05% was obtained. The reaction conditions were optimized and found to be H_2SO_4 (2.4 wt%); reaction temperature, 170°C; and reaction time, 33 min. A high-quality bio-oil was produced that possessed significant ester content. The bio-oil also possessed a low ash content of 0.4% to 0.7%. However, the product had a low pH value (3.8 to 4.0) and thus necessitates storage in acid-resistant bottles (e.g., polypropylene or stainless steel). Thermochemical treatment resulted in a high calorific value of 28.42 MJ kg^{-1}. The bio-oil also had a low nitrogen content compared to bio-oils produced by methods such as pyrolysis or direct liquefaction. A high oxygen content was observed, thus providing the requirement for deoxygenation of the bio-oil. The composition of microalgae bio-oil obtained through thermochemical catalytic liquefaction consists of several methyl and ethyl esters, which result from esterification between organic acids and the glycol solvent, and is similar to that of biodiesel. Campanella et al. (2012) performed thermolysis of microalgae (consisting of mixed wild culture with *Scenedesmus* sp. as the principal constituent) and duckweed (primarily *Wolffia* and *Spirodela* species) in a fixed-bed reactor using CO_2 as a sweep gas for the synthesis of bio-oil and called it "bioleum." The thermolysis of microalgae gave a higher bioleum yield in comparison to that from the duckweed. This is attributed to the difference in composition of the two feedstocks. The fuel properties of the bioleum were found comparable to heavy petroleum crude oil. The use of microalgal oil over lignocellulosic materials for pyrolysis has advantages in the form of lower oxygen concentration and a higher heating value for the former. The heating rate during the thermolysis of microalgae was found to be an important parameter in the formation of the bioleum, where the slow thermolysis did not produce a liquid fuel that could be used as fuel. Maddi et al. (2011) reported that the pyrolysis products of algal (primarily consisting of *Lyngbya* sp. and *Cladophora* sp.) and lignocellulosic biomass (corncobs, woodchips, and rice husk) gave a similar yield of bio-oil. Other compounds that formed along with the bio-oil were bio-char, gases, and ash. The calorific value of lignocellulosic bio-char (except rice husk) was higher than that of algae-derived bio-char. This has been attributed to the higher carbon content in the lignocellulosic biomass. The difference in composition of the bio-oil in the two feedstocks was the presence of nitrogenous compounds in the algal bio-oils. This is assumed to have occurred through degradation of the proteins present in the algae and may decrease its fuel value.

Chakraborty et al. (2012) synthesized bio-oil from *Chlorella sorokiniana* in a two-step sequential hydrothermal liquefaction technique to produce bio-oil and valuable co-products. The bio-oil consisted of 76% carbon, 12% hydrogen, 11% oxygen, 0.78% nitrogen, and 0.16% sulfur. The low nitrogen content avoids the denitrogenation step involved in the production of bio-oil. High oxygen content in the bio-oil necessitates further processing viz. hydrogenation to improve its quality. The yield of the bio-oil obtained by this method consisted of 24% of the dry weight, and optimum polysaccharide extraction occurred at 160°C. The advantage of the two-step sequential hydrothermal liquefaction technique over the direct hydrothermal liquefaction technique was a low formation of bio-char in the former (i.e., 7.6% in comparison to 20.8% in the latter). Li et al. (2012) synthesized bio-oil from the marine brown microalgae, *Sargassum patens* C. Agardh, via hydrothermal liquefaction within a modified reactor. A comparatively moderate yield of 32.1 ± 0.2 wt% bio-oil was obtained in 15 min at 340°C. The feedstock used had a concentration of 15 g biomass per 150 mL water. The bio-oil obtained had a heating value of 27.1 MJ kg^{-1}. The major constituent of the bio-oil was carbon (64.64%), followed by oxygen (22.04%), hydrogen (7.35%), nitrogen (2.45%), and sulfur (0.67%). The characterization of the bio-oil by infrared spectroscopy showed a diverse group of compounds consisting of fats, alkanes, alkenes, alcohols, ketones, aldehydes, carboxylic acids, phenol, esters, ethers, aromatic compounds, nitrogenous compounds, and water. A high concentration of water may be the reason for the low calorific value of the bio-oil produced from the microalgae. Pie et al. (2012) carried out the co-liquefaction of a *Spirulina* and high-density polyethylene (HDPE) mixture in sub- and super-critical ethanol at a reaction temperature of 340°C to obtain bio-oil. The bio-oil thus produced was similar to that obtained from the pure HDPE derived bio-oil. The benefit of the co-liquefaction process of *Spirulina* and HDPE was the synthesis of bio-oil that possessed a high calorific value (48.35 MJ kg^{-1}) due to higher "carbon" and "hydrogen" contents and a lower oxygen content. The samples analyzed by gas chromatograph-mass spectroscopy (GC-MS) showed different compositions for bio-oil derived from *Spirulina*, HDPE, and *Spirulina*–HDPE mixture. While the bio-oil derived from *Spirulina* consisted of oxygen-containing compounds along with fatty acids, fatty acid esters, and ketones as prominent compounds, the bio-oil derived from pure

HDPE consisted of a wide spectrum of hydrocarbons, including saturated and unsaturated aliphatic hydrocarbons. The bio-oil component obtained from the mixture of *Spirulina* and HDPE possessed more hydrocarbons and less oxygen-containing compounds. Hence, the product of the co-liquefaction of *Spirulina* and HDPE was similar in nature to that of pure HDPE liquefaction with a lower reaction temperature needed for thermal degradation of the feedstock. Hu et al. (2012) utilized the microwave-assisted pyrolysis of *Chlorella vulgaris* for the production of bio-oil with a yield of 35.83 and 74.93 wt% using microwave powers of 1,500 and 2,250 W, respectively. It was found that using activated carbon as a catalyst could enhance the bio-fuel yield to 87.47%. The calorific value of the microalgae was determined to be low (21.88 MJ kg^{-1}).

Tabernero et al. (2012) evaluated the industrial potential for production of biodiesel from *Chlorella protothecoides*. It has been estimated that supercritical fluid extraction (supercritical CO_2) for biomass covering a surface area of 7,500 m^2 could generate 10,000 tonnes biodiesel per year in a 150-m^3 bioreactor. Lohrey and Kochergin (2012), in an attempt to minimize the energy consumption of algal biofuels, suggested locating a biodiesel plant close to a sugar mill plant to complement one another. It has been estimated that a cane sugar mill that discards 15% excess bagasse of 10,000 tonnes-per-day capacity can support a 530-ha algae farm to produce 5.8 million L biodiesel per year and will also reduce CO_2 emissions from the mills by 15%. The input in parameters of CO_2, energy, and water are estimated at 2.5 kg kg^{-1}, 3.4 kW-h kg^{-1}, and 1.9 L kg^{-1}, respectively, of algae dry weight.

The fatty acid composition of feedstock plays a significant role in the quality of the biodiesel produced. The European Standard (EN 14214) has limited the linolenic acid (C18:3) content, to not more than 12%. Wu et al. (2012) studied *Chlamydomonas* sp. as a potential feedstock for the synthesis of biodiesel. It was found that *Chlamydomonas* sp. possessed linolenic acid less than 12% and an oleic acid (a monounsaturated fatty acid) constituent of 31.6%. The almost equal compositions of saturated and unsaturated fatty acids in *Chlamydomonas* sp. are desirable for a trade-off between the oxidation stability and low-temperature property of the biodiesel. The FAME (fatty acid methyl ester) content in biodiesel was found to be 25% of total volatile suspended solids from microalgae cultivated using municipal wastewater (Li et al., 2011). Although the ester content in the biodiesel was low, the utilization of microalgae for the production of lipids coupled with wastewater treatment has environmental and economic significance. Upon increasing the ester content in the biodiesel by improving the technology, the process will become far more attractive. Lam and Lee (2012) are of the opinion that biodiesel production will be the ideal product with microalgae as feedstock. To ensure cost effectiveness, the residual biomass after lipid extraction can contain high concentrations of carbohydrates, which should be further utilized for bio-oil and bio-ethanol production. Table 8.1 depicts the ester content and calorific value of the biofuels (biodiesel and bio-oil).

A unique method of thermal analysis to differentiate the oleaginous and non-oleaginous microorganisms (fungi, algae, and yeasts) was developed by Kang et al. (2011). Along with the synthesis of biodiesel, algal biomass residue can be used for other purposes. A linear relationship was observed between exothermic heat and

TABLE 8.1

Comparison of the FAME Ester Content and Calorific Value of the Biodiesel and Bio-oil from Microalgae

Microalgae	Method for Extraction of Oil	Process	Product	FAME Content	HHV/Calorific Value	Ref.
Unidentified	Using hexane as solvent	Transesterification	Biodiesel	Not reported	40 MJ kg^{-1}	11
Chlorella vulgaris (ESP 31)	Sonication to disrupt the cell wall of microalgae and then vigorously mixing the disrupted cell with biphasic solvent of chloroform & methanol	Transesterification	Biodiesel	72.1% (from extracted microalgal oil) 97.25% (upon direct disruption of algal biomass)	—	13
Chlorella sorokiniana	Two-step sequential hydrothermal liquefaction	—	Bio-oil	—	40.8 MJ kg^{-1}	22
Sargassum patens C. Agardh	Hydrothermal liquefaction	—	Bio-oil	—	27.1 MJ kg^{-1}	23
Spirulina/HDPE	Co-liquefaction	—	Bio-oil	—	48.35 MJ kg^{-1}	24
Chlorella vulgaris	Microwave-assisted pyrolysis	—	Bio-oil	—	21.88 MJ kg^{-1a}	25

[a] Low calorific value reported that of microalgae Chlorella vulgaris.

the total lipid content in the tested microorganisms. The exothermic heat per dry sample mass (kJ g^{-1}) in the temperature range from 280°C to 360°C differentiated the oleaginous from the non-oleaginous microorganisms. It was found that the heat evolved from the oleaginous microorganisms was larger than that from the non-oleaginous microorganisms in the specified temperature range. The sharpness of the exothermic peak was also more distinct in the oleaginous microorganisms. Kim et al. (2011) utilized the residual biomass of *Nannochloris oculata* as a biosorbent for the removal of chromium from aqueous solutions. The biological route can also be adopted for biodiesel synthesis. It is anticipated that biodiesel production will increase in the coming years and there will be large amounts of residual biomass that can be used for the treatment of wastewater. The process for the synthesis of biodiesel and bio-oil from microalgae can be depicted through a flowchart (Figure 8.1).

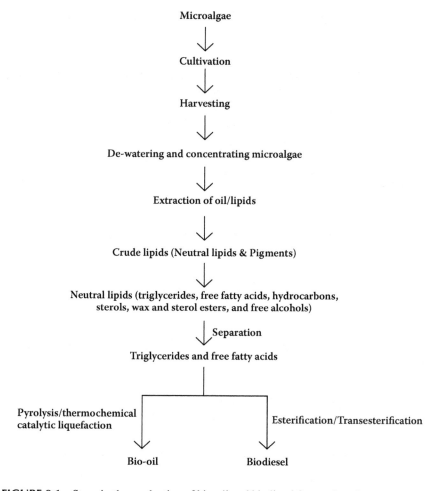

FIGURE 8.1 Steps in the production of bio-oil and biodiesel from microalgae.

8.4 WORLD SCENARIO ON PRODUCTION AND APPLICATION OF BIODIESEL

Biodiesel is now produced in several countries worldwide. The various feedstocks used in the synthesis of biodiesel range from edible to nonedible oils and fats. As per an estimate, Malaysia tops the world in the level of production potential of biodiesel, followed by Indonesia, Argentina, the United States, Brazil, The Netherlands, Germany, Philippines, Belgium, and Spain. The feedstock diverted for the production of biodiesel in these countries is mostly edible oil (28% soybean oil, 22% palm oil, 11% coconut oil, and 5% comprising rapeseed, sunflower, and olive oils). The remaining 20% constitutes animal fats (Sharma and Singh, 2009). However, the developing nations are net importers of edible oil and cannot divert the edible oil for biodiesel production; thus, significant emphasis is placed on alternative feedstocks. Waste cooking and frying oil, waste fat or oil obtained from animal and fish, and microalgal oil have emerged as potential biodiesel feedstocks. Microalgal oil from some species has shown immense potential to produce oil if provided with optimum conditions. The total amount of biodiesel produced worldwide was 16,000 ktonnes in 2009 (Santacesaria et al., 2012). The country that leads in biodiesel production is Germany, followed by France, the United States, Brazil, Argentina, Spain, Italy, Thailand, Belgium, Poland, The Netherlands, Austria, China, Columbia, and South Korea. The biodiesel produced in other countries amounts to 17% of the total production worldwide (http://www.biofuels-platform.ch/en/infos/production.php?id = biodiesel). Biodiesel production from some of the leading countries is listed in Table 8.2. At present, biodiesel is mostly used as transport fuel in

TABLE 8.2
Biodiesel Produced by Leading Countries in 2009

Country	Biodiesel Production (in Ml)	Percentage (%)
Germany	2,859	16
France	2,206	12
United States	2,060	11
Brazil	1,535	9
Argentina	1,340	7
Spain	967	5
Italy	830	5
Thailand	610	3
Belgium	468	3
Poland	374	2
Netherlands	364	2
Austria	349	2
China	338	2
Colombia	330	2
South Korea	300	2
Other countries	2,998	17

compression ignition engines. However, it can also be utilized as fuel to run generator sets, etc.

8.5   BIOFUEL/BIODIESEL FROM MICROALGAL OIL AS A POTENTIAL ALTERNATIVE TO OTHER FUELS

As microalgae possess a simple cellular structure, their capability to efficiently convert solar energy into chemical energy is high. The production of oil per unit area of land from selected microalgae is around 30 times greater than that of terrestrial plants. The scenario thus looks promising for the production of biodiesel from the microalgal oil. There are various steps involved—from the stage of cultivation of microalgae to the final stage of production of bio-oil or biodiesel. The intermediate steps include harvesting, dewatering, concentration, and extraction of microalgal oil. The composition of fatty acids and other constituents present in plant and animal oils varies considerably from microalgal oil. In addition to triglycerides and free fatty acids, microalgal oil contains hydrocarbons, sterols, wax and sterol esters, and free alcohols that cannot be saponified. The major components in microalgae include carbohydrates, proteins, and lipids. In general, the lipid content of microalgal biomass increases when they are deprived of certain nutrients (nitrogen and silicon). However, the deprivation of nitrogen and silicon does not necessarily favor all species viz. *Euglena, Nannochloropsis* strains where cell division has been found to be blocked. There are certain species (such as *Escherichia coli* and *Saccharomyces cerevisiae*) that can be converted to oleaginous species (microbes that can accumulate more than 20% of their cellular dry weight in lipid) by genetic engineering. Although there are several species of microalgae, only a few have been explored with respect to their potential for high biomass yield and lipid content. The FAME content in the biodiesel should have a minimum value of 96.5%, as per the recommendation of EN 14103 (Sarin et al., 2009). However, biodiesel synthesized from only a few of the microalgal species has fulfilled the minimum criteria of ester content in biodiesel as specified by the EN. The reason for this may be attributed to the presence of unsaponifiable constituents in the microalgae. Microalgal species not fulfilling the minimum specified criteria of ester content limits their suitability for biodiesel production. However, microalgal oil can be converted to bio-oil by pyrolysis or thermochemical catalytic liquefaction. The bio-oil can be further upgraded by chemical or physical means. While the chemical upgradation includes processes such as catalytic esterification, catalytic hydroprocessing, and catalytic cracking, the physical upgradation can be done by char removal, hot vapor filtration, liquid filtration, or solvent addition (Xiong et al., 2011). The present status for the production of biodiesel and bio-oil from microalgae is cost intensive. However, the major advantage that the microalgae provide is their growth in aquatic environments and their noncompetitiveness with terrestrial plants. India and many other countries have a vast coastal area where microalgae can be grown, cultured, and harvested. The other important benefit of microalgae is the suitability of some of the species in wastewater. Thus, microalgae cultured with wastewater will have the dual benefit: of the production of oil and the treatment/disposal of wastewater. Numerous strains of microalgae are available in nature, and several of these species may be explored for their feasibility to be cultured as oleaginous species. The future seems bright for microalgae to provide a future alternative to the other fuels.

8.6 CONCLUSION

The oil extracted from microalgae consists of polar lipids and neutral lipids. The neutral lipids consist of triglycerides, free fatty acids, hydrocarbons, sterols, wax and sterol esters, and free alcohols. Because only triglycerides and free fatty acids are saponifiable, they must be separated from the others so as to convert them to biodiesel by esterification or transesterification. The synthesis of biodiesel/bio-oil from microalgae involves several steps, including selection of an appropriate species among a large diversity of species of microalgae (around 300,000). For the production of bio-oil, a thermochemical method is adopted for the preparation of fuel from microalgae through pyrolysis, direct combustion, or thermochemical liquefaction, wherein the organic compound is thermally decomposed at high temperature in the absence of oxygen. A high yield of bio-oil (97.05%) was obtained through liquefaction of *Dunaliella tertiolecta*. Thermochemical catalytic liquefaction has an advantage over pyrolysis or direct liquefaction, in that a low nitrogen content is present. A high oxygen content has been observed, which requires deoxygenation of the bio-oil. Other compounds are also formed along with bio-oil, such as bio-char, gases, and ash, all of which lower the calorific value of the fuel. The amount of biodiesel obtained from microalgal oil can be enormous The fatty acid composition of the feedstock has been found to play a significant role in the composition of the biodiesel. A trade-off between the oxidation stability and low-temperature properties of biodiesel has been observed and hence a balance between the two must be maintained.

ACKNOWLEDGMENTS

Bhaskar Singh is grateful to the Council of Scientific and Industrial Research (CSIR) New Delhi, India for the award of Research Associateship.

REFERENCES

Batan, L., Quinn, J., Willson, B., and Bradley, T. (2010). Net energy and greenhouse gas emission evaluation of biodiesel derived from microalgae. *Environmental Science and Technology*, 44: 7975–7980.

Campanella, A., Muncrief, R., Harold, M.P., Griffith, D.C., Whitton, N.M., and Weber, R.S. (2012). Thermolysis of microalgae and duckweed in a CO_2-swept fixed-bed reactor: Bio-oil yield and compositional effects. *Bioresource Technology*, 109: 154–162.

Chakraborty, M., Miao, C., McDonald, A., and Chen, S. (2012). Concomitant extraction of bio-oil and value added polysaccharides from *Chlorella sorokiniana* using a unique sequential hydrothermal extraction technology. *Fuel*, 95: 63–70.

Chen, Y.H., Huang, B.Y., Chiang, T.H., and Tang, T.C. (2012). Fuel properties of microalgae (*Chlorella protothecoides*) oil biodiesel and its blends with petroleum diesel. *Fuel*, 94: 270–273.

Chowdhury, R., Viamajala, S., and Gerlach, R. (2012). Reduction of environmental and energy footprint of microalgal biodiesel production through material and energy integration. *Bioresource Technology*, 108: 102–111. http://www.biofuels-platform.ch/en/infos/ production.php?id=biodiesel.

Hu, Z., Ma, X., and Chen, C. (2012). A study on experimental characteristics of microwave-assisted pyrolysis of microalgae. *Bioresource Technology*, 107: 487–493.

Johnson, M.B., and Wen, Z. (2009). Production of biodiesel fuel from the microalga *Schizochytrium limacinum* by direct transesterification of algal biomass. *Energy and Fuels*, 23: 5179–5183.

Kang, B., Honda, K., Okano, K., Aki, T., Omasa, T., and Ohtake, H. (2011). Thermal analysis for differentiating between oleaginous and non-oleaginous microorganisms. *Biochemical Engineering Journal*, 57: 23–29.

Kao, C.Y., Chiu, S.Y., Huang, T.T., Dai, L., Wang, G.H., Tseng, C.P., Chen, C.H., and Lin C.S. (2012). A mutant strain of microalga *Chlorella* sp. for the carbon dioxide capture from biogas. *Biomass and Bioenergy*, 36: 132–140.

Kim, E.J., Park, S., Hong, H.J., Choi, Y.E., and Yang, J.W. (2011). Biosorption of chromium (Cr(III)/Cr(VI)) on the residual microalga *Nannochloris oculata* after lipid extraction for biodiesel production. *Bioresource Technology*, 102: 11155–11160.

Lam, M.K., and Lee, K.T. (2012). Microalgae biofuels: A critical review of issues, problems and the way forward. *Biotechnology Advances*, 30: 673–690.

Lardon, L., Helias, A., Sialve, B., Steyer, J.P., and Bernard, O. (2009). Life-cycle assessment of biodiesel production from microalgae. *Environmental Science and Technology*, 44: 6475–6481.

Li, D., Chen, L., Xu, D., Zhang, X., Ye, N., Chen, F., and Chen, S. (2012). Preparation and characteristics of bio-oil from the marine brown alga *Sargassum patens* C. Agardh. *Bioresource Technology*, 104: 737–742.

Li, Y., Zhou, W., Hu, B., Min, M., Chen, P., and Ruan R.R. (2011). Integration of algae cultivation as biodiesel production feedstock with municipal wastewater treatment: Strains screening and significance evaluation of environmental factors. *Bioresource Technology*, 102: 10861–10867.

Liu, X., Clarens, A.F., and Colosi, L.M. (2012). Algae biodiesel has potential despite inconclusive results to date. *Bioresource Technology*, 104: 803–806.

Lohrey, C., and Kochergin, V. (2012) Biodiesel production from microalgae: Co-location with sugar mills. *Bioresource Technology*, 108: 76–82.

Maddi, B., Viamajala, S., and Varanasi, S. (2011). Comparative study of pyrolysis of algal biomass from natural lake blooms with lignocellulosic biomass. *Bioresource Technology*, 102: 11018–11026.

Pie, X., Yuan, X., Zeng, G., Huang, H., Wang, J., Li, H., and Zhu, H. (2012). Co-liquefaction of microalgae and synthetic polymer mixture in sub- and supercritical ethanol. *Fuel Processing Technology*, 93: 35–44.

Rosch, C., Skarka, J., and Wegerer, N. (2012) Materials flow modeling of nutrient recycling in biodiesel production from microalgae. *Bioresource Technology*, 107: 191–199.

Santacesaria, E., Vicente, G.M., Serio, M.D., and Tesser, R. (2012). Main technologies in biodiesel production: State of the art and future challenges. *Catalysis Today*, 195(1): 2–13.

Sarin, R., Kumar, R., Srivastav, B., Puri, S.K., Tuli, D.K., Malhotra, R.K., and Kumar, A. (2009). Biodiesel surrogates: Achieving performance demands. *Bioresource Technology*, 100: 3022–3028.

Sharma, Y.C., and Singh, B. (2009). Development of biodiesel: current scenario. *Renewable and Sustainable Energy Reviews*, 13: 1646–1651.

Sharma, Y.C., Singh, B., and Korstad, J. (2011). A critical review on recent methods used for economically viable and eco-friendly development of microalgae as a potential feedstock for synthesis of biodiesel. *Green Chemistry*, 13: 2993–3006.

Siegler, H.D.H., McCaffrey, W.C., Burrell, R.E., and Zvi, A.B. (2012). Optimization of micro-
    algal productivity using an adaptive, non-linear model based strategy. *Bioresource
    Technology*, 104: 537–546.
Sivakumar, G., Xu, J., Thompson, R.W., Yang, Y., Smith, P.R., and Weathers, P.J. (2012).
    Integrated green algal technology for bioremediation and biofuel. *Bioresource
    Technology*, 107: 1–9.
Stephenson, P.G., Moore, C.M., Terry, M.J., Zubkov, M.V., and Bibby, T.S. (2011). Improving
    photosynthesis for algal biofuels: Toward a green revolution. *Trends in Biotechnology*,
    29: 615–623.
Tabernero, A., del Valle, E.M.M., and Galan, M.A. (2012). Evaluating the industrial potential
    of biodiesel from a microalgae heterotrophic culture: Scale-up and economics.
    Biochemical Engineering Journal, 63: 104–115.
Tran, D.T., Yeh, K.L., Chen, C.L., and Chang, J.S. (2012). Enzymatic transesterification of
    microalgal oil from *Chlorella vulgaris* ESP-31 for biodiesel synthesis using immobilized
    Burkholderia lipase. *Bioresource Technology*, 108: 119–127.
Velasquez-Orta, S.B., Lee, J.G.M., and Harvey, A. (2012). Alkaline in situ transesterification
    of *Chlorella vulgaris*. *Fuel*, 94: 544–550.
Vijayaraghavan, K., and Hemanathan, K. (2009). Biodiesel production from freshwater algae.
    Energy and Fuels, 23: 5448–5453.
Wu, L.F., Chen, P.C., Huang, A.P., and Lee, C.M. (2012). The feasibility of biodiesel production
    by microalgae using industrial wastewater. *Bioresource Technology*, 113: 14–18.
Xiong, W.M., Fu, Y., Zeng, F.X., and Guo, Q.X. (2011). An in situ reduction approach for bio-
    oil hydroprocessing. *Fuel Processing Technology*, 92: 1599–1605.
Zou, S., Wu, Y., Yang, M., Li, C., and Tong, J. (2009). Thermochemical catalytic liquefaction of
    the marine microalgae *Dunaliella tertiolecta* and characterization of bio-oils. *Energy and
    Fuels*, 23: 3753–3758.

9 Analysis of Microalgal Biorefineries for Bioenergy from an Environmental and Economic Perspective Focus on Algal Biodiesel

Susan T.L. Harrison, Christine Richardson, and Melinda J. Griffiths
Centre for Bioprocess Engineering Research
Department of Chemical Engineering
University of Cape Town, South Africa

CONTENTS

9.1   MICROALGAE FOR BIOENERGY

Microalgae are well recognized for their potential to contribute as an important energy source as well as to provide a renewable feedstock for commodity organic products. Their role as a renewable energy source has promise for both the production of liquid fuels, including diesel, gasoline, and jet fuels, and electricity generation. Routes to these energy sources include accumulation of algal oil for transesterification to biodiesel, fermentation to alcohols for inclusion in gasoline, production of hydrogen, anaerobic digestion to biogas, thermal processing to bio-oil, co-combustion, and gasification (Chisti and Yan, 2011).

The technical feasibility of algal biomass as a source of bioenergy has been demonstrated for a number of the products described above. Further evidence of the commercial interest in these products is demonstrated by both the R&D investment of leading energy companies, including Exxon, BP, Chevron, Shell, and Neste Oil (Norsker et al., 2011), as well as a number of start-up companies attempting commercialization of algal fuels (Table 9.1). However, it is recognized that these products remain expensive in comparison to petroleum-based products. Further, it is essential to understand the environmental benefits of these process routes objectively. Optimization of the product spectrum in terms of economic and environmental sustainability is best informed by a rigorous analytical approach that informs the key process targets for their improvement.

This chapter focuses on the economic and environmental impacts of the production of algal oil and biodiesel as primary products, owing to the comprehensive analysis of these routes. Processing of algal biomass to secondary products in a biorefinery approach is included to explore the potential of algal energy more completely. The alternative bioenergy products are not considered comprehensively here.

TABLE 9.1
Start-Up Companies for Algal Biofuels

Company	Location	Website
Algenol Biofuels	Bonita Springs, FL, USA	www.algenolbiofuels.com
Aquaflow	Nelson, New Zealand	www.aquaflowgroup.com
Aurora Algae Inc	Hayward, CA, USA	www.aurorainc.com
Bioalgene	Seattle, WA, USA	www.bioalgne.com
Bionavitas, Inc.	Redmond, WA, USA	www.bionavitas.com
Bodega Algae, LLC	Boston, MA, USA	www.bodegaalgae.com
LiveFuels, Inc.	San Carlos, CA, USA	www.livefuels.com
PetroAlgae Inc.	Melbourne, FL, USA	www.petroalgae.com
Phyco Biosciences	Chandler, AZ, USA	www.phyco.net
Sapphire Energy, Inc.	San Diego, CA, USA	www.sapphireenergy.com
Seambiotic Ltd.	Tel Aviv, Israel	www.seambiotic.com
Solazyme, Inc.	South San Francisco, CA, USA	www.solazyme.com
Solix Biofuels, Inc.	Fort Collins, CO, USA	www.solixbiofuels.com
Synthetic Genomics, Inc.	La Jolla, CA, USA	www.syntheticgenomics.com

Source: From Chisti and Yan (2011).

9.2 ANALYTICAL TOOLS FOR ASSESSING ENVIRONMENTAL SUSTAINABILITY

Over the past two decades, much work has focused on methodology to assess the environmental impact of processes and products. A number of these approaches are summarized in Table 9.2, indicating the methodology and nature of the assessment. It must be noted that while initially bioprocesses and energy processes from renewable resources were assumed to be preferential with respect to lower

TABLE 9.2
Approaches to the Quantification of Environmental Sustainability of Process Options

Tool	Approach	Ref.
Environmental Impact Assessment (EIA)	"The process of identifying, predicting, evaluating and mitigating the biophysical, social, and other relevant effects of development proposals prior to major decisions being taken and commitments made."	International Association for Impact Assessment (IAIA), 1999
	A shortcoming is the narrow spatial and temporal scope, typically limited to the site of the project.	
Eco-efficiency	Centered on producing cost-effective goods and services while reducing their environmental impact; i.e., "producing more with less."	
Ecological footprinting	A measure of the demand placed on the Earth's resources through human activity. This is developed in terms of the biologically productive area (land and sea) required to produce the materials used and to assimilate the wastes produced. Developing consistency in the methods used to calculate ecological footprint is currently a key focus.	Wackernagal et al., 2002
Carbon footprinting (ISO 14064)	"A measure of the total amount of carbon dioxide (CO_2) and methane (CH_4) emissions of a defined population, system or activity, considering all relevant sources, sinks and storage within the spatial and temporal boundary of the population, system or activity of interest. Calculated as carbon dioxide equivalent (CO_2e) using the relevant 100-year global warming potential (GWP100)."	UK Carbon Trust, 2008
Life Cycle Assessment (LCA) (ISO 14040 series)	Analytical tool to assess environmental impacts of processes through definition of goal and scope, inventory analysis, impact assessment, interpretation. Uses assessment software packages including SimaPro™, Umberto®, GaBi™, and TEAM™.	Consoli et al., 1993; ISO 14044, 2006
	Key advantages of this approach include that it is not location specific, allows comparison across processes, and is built on a strong literature database.	
Net energy recovery (NER)	NER = energy produced/energy input	

environmental impact, it has been demonstrated clearly that this does not necessarily hold; hence, objective assessment of the environmental burden of each process is essential in product and process selection, in a similar manner to that used to ensure economic feasibility.

Life cycle assessment (LCA) systematically identifies environmental impact and opportunities to minimize it, and evaluates these (Curran 2000). It is supported by a strong literature database and a well-defined methodology. A track record exists for its use in the environmental assessment of biofuels (Kaltschmitt et al., 1997; Kim and Dale, 2005; von Blottnitz and Curran, 2007; Harding et al., 2008; Evans et al., 2009). In conducting the LCA, setting the goal and scope of the study allows for selecting a functional unit for comparison and setting the system boundaries. A full inventory of the process flowsheet is required, including all raw materials and energy, and all emissions and products generated. Data are preferably obtained from operating plants; where this is not feasible in new process development, data are obtained experimentally, from the literature or through modeling, and validated through material and energy balancing. Typically, a cradle-to-gate approach is used where the products formed are the same. Where the products formed differ from the existing product and result in different emissions and by-products on use, a cradle-to-grave approach is needed to consider product use and disposal. In both cases, the raw material and energy requirements are expanded to include their pre-processing, taking into account extraction from abiotic reserves, cultivation, agricultural processes, etc. Typically, the impact of construction of the process plant and equipment is negligible with respect to the impact of the operating plant. In new technology environments, this should be verified. This has been demonstrated for algal biodiesel in all categories except land use (Lardon et al., 2009). Where reactors having a short life span are used (e.g., polyethylene bags or PVC linings), these need to be included in the analysis. For multiple products or by-products, as in the biorefinery, environmental burden allocation or substitution is required to allocate the overall burden representatively across the products formed. Burden allocation may be done based on the mass or volume ratio of useful products or, in some cases, based on cost. According to ISO (International Organization for Standardization) guidelines, substitution is preferred where possible; that is, the additional product or by-product is accounted for through the inventory typical of its conventional process route. This handling of multiple products is important as typically the production of multiple biofuels has been shown to increase the material and energy efficiency and process economics of biomass utilization (Kaparaju et al., 2009).

Life cycle inventory (LCI) data are used in life cycle impact assessments (LCIAs), typically using appropriate software to group the impacts into a manageable set of impact categories (mid-point categories), such as abiotic depletion, global warming, eutrophication, acidification, toxicity, etc. These may be further grouped into end-point categories, such as human health, climate change, and ecosystem quality, where appropriate.

The importance of the holistic study, considering all aspects of resource utilization and emission generation, is demonstrated through early-stage biofuel analyses where the carbon benefits of land use were counted for first-generation biofuels; however, the emissions caused by clearing of the land to grow new feedstock (land-use change) were not estimated (Searchinger, 2008). Fargione (2008) determined that the greenhouse gases (GHGs) released from changing natural habitats to biofuel cropland were several-fold greater than the offset from displacing fossil fuels, and hence a "carbon payback time" was defined to determine the time required before a true reduction in GHG resulted. This example drives home the need for an integrated assessment of environmental impacts.

9.3 ENVIRONMENTAL SUSTAINABILITY OF MICROALGAL PROCESSES

9.3.1 OVERVIEW OF ENVIRONMENTAL ASSESSMENT OF ALGAL BIODIESEL

While prior studies on biofuels were largely limited to feedstocks of terrestrial plant origin, over the past 4 years, a number of studies have been published on the environmental analysis of algal energy processes (Lardon et al., 2009; Batan et al., 2010; Clarens et al., 2010; Jorquera et al., 2010; Sander and Murthy, 2010; Stephenson et al., 2010; Razon and Tan, 2011; Richardson et al., 2012a). Prior to this, environmental analyses of algal energy processes were limited to the work of Kadam (2002) and Sazdanoff (2006). The former considered the benefit of co-combustion of coal and algae in electricity generation, while the latter presented a model for the algal-to-biodiesel fuel cycle, including climatic data to simulate production at four locations in the United States. In most studies, analyses of the energy and global warming potential (GWP) have formed the key assessment criteria, with net energy recovery (NER) and LCA being the most frequently used approaches. In all cases, the absence of commercial-scale inventory data implies that scale-up estimates from laboratory- and pilot-scale data inform these analyses, requiring that a range of assumptions must be made on large-scale performance within these systems. In Table 9.3, the systems analyzed in each of the studies reported are summarized. These can be positioned within the context of Figure 9.1, which provides a block flowsheet of the integrated process for the production of biodiesel from microalgae and demonstrates the manner in which different studies focus on different components of the process. The findings in these studies are discussed in the following sections.

It must be emphasized that the immature nature of the microalgal biofuels process implies that the data have mostly been sourced from laboratory studies, modeling, and some pilot-scale research. Variability in data is a source of variation in results presented in the literature. As an example, Collet et al. (2010) estimated the energy for paddlewheels and pump of water at 0.2 and 0.153 kW-h per kilogram algae, respectively, whereas Clarens et al. (2010) suggested values of 0.035 and 0.029 kW-h per kilogram algae, respectively. Substitution with the lower values assumed by Clarens et al. (2010) leads to a 44% reduction in total energy demand. Further, lipid productivity has been recognized as a key factor in selecting conditions for biofuel production (Griffiths and Harrison, 2009; Rodolfi et al., 2009). Lipid productivity is the product of lipid content and specific growth rate. It is recognized that nutrient limitation results in compromised growth rates and high lipid content; hence, care must be exercised to utilize compatible data when estimating lipid productivity. Examples are found in the literature where the high specific growth

TABLE 9.3

Studies of the Environmental Benefit or Burden of Processes for Algal Biofuels

Investigators	Algal Species; Primary Product	Objective and Process	System Components and Boundaries	Reactor System	Environmental Analysis	Source of Data
Lardon et al., 2009	*Chlorella vulgaris*; biodiesel	To assess the environmental impact of biodiesel from microalgae as technologically immature process and to identify barriers to large-scale environmental sustainability. Nutrient supply was compared with nutrient starvation; wet and dry extraction was tested; AD and nutrient recycle were used	Infrastructure, cultivation, dewatering, extract & convert, transport & distribute, combustion.	Open raceways	LCA of virtual facility, using CML. Allocation based on energetics	Laboratory data
Batan et al., 2010	*Nannochloropsis salina*; Microalgal biodiesel	To evaluate an industrial-scale model for microalgal biodiesel production in photobioreactors and compare microalgal biodiesel with soybean-based diesel and petroleum diesel	Cultivation, dewatering, AD, extract & convert, transport & distribute.	Polyethylene PBR bags	GREET model, NER, LCA based on GHG, allocation by substitution	Colorado State University pilot plant

Clarens et al., 2010	Generalized microalgal system	To compare algae to switchgrass, canola, and corn as a bioenergy feedstock, focusing on algal production; effect of wastewater as nutrient source and flue gas are considered	Infrastructure, cultivation, dewatering	Open raceway ponds; flocculation and centrifugation	LCA, no allocation	Laboratory data; meteorological data
Campbell et al., 2010	Species not specified, Biodiesel	To investigate environmental impact and economic feasibility of biodiesel production in Australia, using three levels of CO_2 supply and two growth rates in ponds	Cultivation, dewatering, AD, extract & convert, transport & distribute, combustion	Raceway pond	LCA - GHG, no allocation	
Jorquera et al., 2010	*Nannochloropsis* sp.; oil-rich algal biomass	To investigate the production of algal biomass in raceway ponds, tubular and flat-plate photobioreactors to evaluate feasibility	Infrastructure, cultivation	Raceway ponds; tubular reactors; flat-plate reactors	NER; energy LCA, no allocation	Laboratory-scale data from literature
Sander and Murthy, 2010	Algal biodiesel	Through investigating the process sustainability and net energy balance of the algal biodiesel process, provide baseline information for process development; integration with carbohydrate conversion to ethanol	Infrastructure, cultivation, dewatering, AD, extract & convert, transport & distribute	Inoculum in photobioreactors or indoor ponds, open ponds for process		USLCI database

(Continued)

TABLE 9.3 (*Continued*)
Studies of the Environmental Benefit or Burden of Processes for Algal Biofuels

Investigators	Algal Species; Primary Product	Objective and Process	System Components and Boundaries	Reactor System	Environmental Analysis	Source of Data
Stephenson et al., 2010	*Chlorella vulgaris*; Biodiesel	To compare open raceways and tubular photobioreactors for algal biomass production in a two-stage process with nitrogen limitation (phase 2) for lipid accumulation; considered a range of DSP options, including AD of residual biomass	Infrastructure, cultivation, dewatering, AD, extract & convert, transport & distribute, combustion	Raceways; airlift tubular reactors	LCA using EDIP, market value allocation and substitution	Laboratory- (own) and pilot-scale (lit.) data used (Rodolfi et al., 2009)
Collet et al., 2010	*Chlorella vulgaris*, biogas, biodiesel	To assess the environmental impact of biogas generation from microalgae, with nutrient recycle, and compare this to microalgal biodiesel	Infrastructure, cultivation, dewatering, AD	Raceway pond	LCA using CML, allocation by substitution	
Yang et al., 2011	*Chlorella vulgaris*	To determine the water footprint and nutrient usage of microalgal biodiesel production, considering use of wastewater, nutrient recycle	Cultivation, dewatering, AD, extract & convert, transport & distribute	Raceway pond	Water, no allocation	

Reference	Species; products	Objective	Stages	Reactor	Analysis	Data source
Razon and Tan, 2011	*Haematococcus pluvialis* and *Nannochloropsis* sp.; biodiesel and biogas	To determine whether microalgal biodiesel can deliver more energy than is required to produce it; the analysis considers the integration of biogas, the use of wastewater, and nutrient recycle and wet extraction	Cultivation, dewatering, AD, extract & convert	Photobioreactor inoculum; raceway pond; gravitational settling	NER	Literature data
Richardson et al., 2012	*Phaeodactylum tricornutum*	To consider the impact of reactor selection and operating conditions on energy requirement and, more broadly, environmental impact	Cultivation, dewatering, AD, extract & convert	Tubular airlift reactor, horizontal tubular reactor, raceway	NER and LCA	Data from ASP (Sheehan et al., 1998)
Richardson, 2011	*Chlorella vulgaris*, *Tetraselmus*, *Scenedesmus* sp.	To understand the impact of algal species selection on the environmental impact of the algal biorefinery	Cultivation, dewatering, AD, extract & convert	Tubular airlift reactor	NER and LCA	Own data from lab

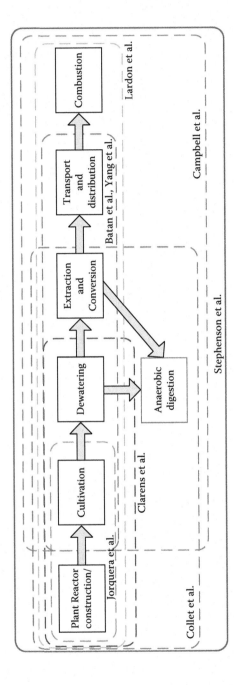

FIGURE 9.1 Block flow diagram of the integrated microalgal biodiesel production process, indicating system boundaries of a selection of LCA studies.

rates of nutrient-replete growth are combined with high lipid content of nitrogen limitation to calculate productivity—such productivities are typically unobtainable. These examples highlight the importance of data quality and the challenges found owing to the absence of reliable production data. Clearly, sensitivity analysis forms a key component of these studies. Such sensitivity analyses are presented by, among others, Stephenson et al. (2010), Richardson et al. (2012b), Williams and Laurens (2010), and Norsker et al. (2011).

9.3.2 Algal Bioreactors

In most studies performed to consider the environmental burden of microalgal biodiesel, the key contributors to the algal cultivation process have been identi-fied across raceway ponds and closed photobioreactors. Across all reactor configu-rations, the major contributions to environmental burden in terms of net energy ratio, abiotic depletion, and GHGs were incurred from the energy requirement for mass transfer and mixing in the reactor, as well as the energy requirement asso-ciated with the provision of combined nitrogen for cell growth (Lardon et al., 2009; Batan et al., 2010; Stephenson et al., 2010; Richardson et al., 2012b). These requirements are sensitive to algal biomass concentration, lipid content, and algal productivity. Algal productivity and concentration were the most influential (e.g., Stephenson et al., 2010; Razon and Tan, 2011), owing to their impact on system volume, influencing both energy input for mixing and pumping, and the amounts of nutrients required, as previously demonstrated in other microbial systems (Harding et al., 2008; Harding et al., 2012).

Studies by Jorquera et al. (2010), Stephenson et al. (2010), and Richardson et al. (2012b) compared selected types of reactors. Jorquera et al. (2010) compared horizontal tubular reactors, flat-plate reactors, and raceway ponds using a basis of 100 tons algal biomass per year. The closed photobioreactors provided higher biomass concentrations, and higher volumetric and areal productivities than ponds. Correlated with this, the raceway ponds had an increased land requirement. However, their energetic requirements were significantly lower than the closed photobioreac-tors. Under their operating conditions, the NER of the horizontal tubular reactor illustrated that it was not feasible in terms of either oil or biomass production (NER of 0.07 and 0.20, respectively). The NER of the flat-plate reactor was 54% of that for the open raceway for oil production (NER of 1.65 and 3.05, respectively) and for production of algal biomass (NER of 4.51 and 8.34, respectively).

Stephenson et al. (2010) compared the performance of the integrated algal biodiesel process, from cradle to combustion, using the open raceway and tubular airlift photobioreactor and a two-stage cultivation method to maximize lipid for-mation under nitrogen starvation in the second stage. In their system, the reactor was the dominant contributor to energy consumption. The design of the tubular airlift photobioreactor required energy input an order of magnitude greater than the raceway on an energy equivalence basis, despite its higher productivity. While 85% of the energy requirement of the tubular reactor was attributed to operation and the remainder to reactor manufacture, the latter exceeded the total GWP of the raceway system.

Similarly, Richardson et al. (2012b) compared the performance of the raceway, horizontal tubular reactor, and airlift tubular reactor as a component of the algal biorefinery producing biodiesel and biogas. Their comparison was made using literature data for *Phaeodactylum tricornutum*, extensively studied in the Aquatic Species Programme (Sheehan et al., 1998). Considering an integrated biorefinery system with combined biogas production, the relative NER values for these photobioreactors compared to the raceway were 64% and 8%, respectively. Under the operating conditions selected, the NER of the airlift reactor was unacceptable, owing to the low productivity achieved relative to the energy input. The airlift reactor can be operated at much reduced gas flow rates and concomitant reduction in energy input without compromising productivity (data not shown), indicating the need to make these comparisons using optimized performance data relevant to commercial-scale operation. In all cases, the reactor energy requirement dominated that of the process, with that of the horizontal tubular reactor being some twofold that of the raceway per unit biodiesel. Owing to the lower biomass and oil concentrations achieved in the raceway reactor, this advantage of reduced reactor energy was partially offset by the greater pumping energy required for the larger volume processed from the raceway (2.2-fold); the pumping energy within the raceway biorefinery is a quarter of the reactor energy requirement. Extending the analysis beyond energy, the acidification and eutrophication impacts of the horizontal tubular reactor were 61% and 73%, respectively, of the raceway system under the standard operating conditions selected. The GWP was negative for the raceway system compared to a positive value of 60% for the horizontal tubular reactor.

Recognizing the increased energy requirement of traditional photobioreactors for mixing and mass transfer as well as manufacture, Batan et al. (2010) assessed the sparged polyethylene photobioreactor bags. While they report positive NER values for these systems, agreement in the literature with respect to the feasibility of the airlift system for biofuel production has not been found and further assessment is required.

Razon and Tan (2011) assessed the combined production of biodiesel and biogas using *Haematococcus pluvialis*. While low biomass concentrations and productivity resulted in a negative NER, the energy requirement of the flat-plate bioreactor used for intermittent inoculum supply exceeded that of the raceway system used for large-scale production by some twofold, thus confirming the much higher energy demand per unit biofuel in closed reactor systems.

Stephenson et al. (2010) disaggregated the contributions to the reactor energy and GWP in the tubular airlift reactor and raceway pond. These are shown relatively in Figure 9.2 (see color insert), where the total fossil energy requirements estimated for cultivation in the tubular airlift reactor and raceway under standard conditions were approximately 230 and 29 GJ per tonne biodiesel formed, respectively. The corresponding GWPs were 13,550 and 1,900 kg CO_2 per tonne biodiesel, respectively. In addition to the magnitude, the relative values illustrate that electrical energy for mixing and mass transfer dominates the reactor energy and related GWP in the tubular reactor. The lower mixing energy in the raceway results in the energy for combined nitrogen provision being significant. Furthermore, the much larger raceways required, owing to lower productivities achieved, lead to the construction components making a more dominant contribution, especially the PVC liners of the ponds.

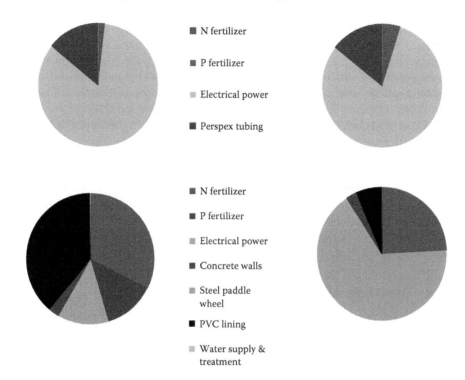

N fertilizer

P fertilizer

Electrical power

Perspex tubing

N fertilizer

P fertilizer

Electrical power

Concrete walls

Steel paddle wheel

PVC lining

Water supply & treatment

FIGURE 9.2 (See color insert.) The relative contribution of fossil energy (left) and GWP (right) to the total requirements for microalgal biodiesel production using a tubular airlift reactor (upper) and a raceway (lower). From the LCA of *C. vulgaris* conducted by Stephenson et al. (2010) under standard conditions. The total fossil energy requirements of 230 and 29 GJ and GWP of 13,550 and 1,900 kg CO_2 per tonne biodiesel formed were estimated for the tubular reactor and raceway, respectively.

9.3.3 NUTRIENT PROVISION FOR MICROALGAL CULTURE

CO_2 provision to raceway ponds through additional sparging of compressed CO_2 contributed some 40% of the energy consumption and 30% GHG emissions (Clarens et al., 2010). Alternatively, co-location with industries or facilities with high CO_2 emissions (such as power stations, fermentation plants, or anaerobic digester systems) may facilitate reduced contributions for effective CO_2 provision. This has been demonstrated; however, it is noted that the CO_2 source may influence the productivity achieved and interacts with the supply of other nutrients. Stephenson et al. (2010) modeled the impact of CO_2 concentration in the gas to be compressed for sparging into either the raceway or tubular reactor system. Owing to the influence of CO_2 concentration on both concentration driving force (and hence transfer rate) and on the volume of gas to be compressed, its impact is significant, with the fossil energy requirement nearly doubling on decreasing the CO_2 concentration from 12.5% (typical of flue gases) to 9% and increasing fourfold on decrease to 5% by volume in the raceway system. The design of a low-depth carbonation sump also favors reduced energy consumption.

It is increasingly recognized that the provision of nutrients, especially combined nitrogen, to bioprocesses affects their life-cycle impact (Harding 2009; Harding et al., 2012). Lardon et al. (2009) illustrated that the provision of fertilizer accounted for 15% to 25% of the energy requirements per unit biodiesel. This was substantially reduced under nitrogen-limited conditions (6% to 9%). Clarens et al. (2010) illustrated that 50% of the energy requirement and GWP for biomass production is attributable to nutrient provision in their raceway system. The potential exists to replace these with wastewaters such as effluent from the conventional activated sludge process or source-separate urine. The former has the potential to provide a water supply simultaneously. Similar benefits can be achieved by maximizing the nutrient recycle (Stephenson et al., 2010; Richardson, 2011) and minimizing the nitrogen input required, either by optimizing the nitrogen limitation or selecting an algal species of low nitrogen content, for example, *Phaeodactylum tricornutm* at 0.8% N over algal species with a typical nitrogen content of 6% by mass (Richardson et al., 2012b).

9.3.4 BIOMASS RECOVERY

Dewatering of the dilute algal suspensions required to minimize light limitation in the bioreactor is recognized as a key challenge in large-scale algal processes. This is aggravated by small algal cell size and a density only slightly greater than that of water. An initial dewatering step is required to achieve a solids concentration of 1% to 2.5% by mass prior to concentration using an energy-intensive operation such as centrifugation to a solids concentration of 5% to 20% (Benemann and Oswald, 1996). Ideally, flocculation and sedimentation are used for the primary dewatering, facilitated by the algal species selected (Lardon 2009; Stephenson et al., 2010). Other options include natural settling (Collet et al., 2010), filtration (Yang et al., 2011), and dissolved air flotation (Campbell et al., 2010). The decanter centrifuge, spiral-plate centrifuge, and rotary press are most typically proposed for the second dewatering step (Lardon et al., 2009; Campbell et al., 2010; Clarens et al., 2010; Collet et al., 2010; Stephenson et al., 2010). Stephenson et al. (2010) report a relative electricity consumption for flocculation of 1 unit versus 4 units for centrifugation and 14.4 units for the cultivation. On using a more dilute algal culture, as typically found with the raceway, both pumping energy and the need to increase flocculant supplied to maintain its volumetric concentration impact the process.

9.3.5 BIOMASS DRYING AND CONVERSION

In a number of studies, it has been assumed that the biodiesel production from algal oil should be conducted using the same approach as for vegetable oil. This requires the drying of the algal biomass to 90% solids by mass (Sazdanoff 2006; Lardon et al., 2009; Batan et al., 2010; Yang et al., 2011). Typically, belt or drum drying is used, with energy provided by natural gas. This may represent the major or a significant portion of the energy demand of the process. Sander and Murthy (2010) estimated that 89% of the energy requirement of their "well-to-pump," raceway-based algal biodiesel system was required for drying the intermediate product prior to conversion to biodiesel, using a natural gas powered dryer. Razon and Tan (2011) attributed 48% of their energy requirement (310 MJ per tonne biodiesel) to drying, following production of 1 kg FAME and 1.5 m^3 biogas using a raceway pond. Drying of the biomass is only feasible where natural resources (e.g., solar drying) can be used while preventing lipid oxidation (Lardon et al., 2009).

Furthermore, the requirement of cell disruption for product recovery has also been considered. Razon and Tan (2011) demonstrated that the energy input to the bead mill for disruption of *Haematococcus pluvialis* formed the greatest contribution to the process energy required (>30%). Stephenson et al. (2010) estimated cell disruption by high-pressure homogenization to account for approximately 5 GJ tonne^{-1} biodiesel formed and 320 kg CO_2 eq tonne^{-1} biodiesel in GWP. The need for cell disruption is a function of both the flowsheet selected and the algal species.

In a comparison of the hypothetical wet and dry processing routes for the transesterification to biodiesel, the benefit of developing an effective wet processing route for which a discrete cell disruption step and rigorous drying are not essential is clearly demonstrated (Lardon et al., 2009). The energy requirement was reduced to 70% to 75% of the dry processing route while the energy recoverable through further processing of the oil cake increased by 67% to 115%. Overall, the additional energy recoverable was 0.6 to 1.1 MJ per 1 MJ biodiesel. Direct esterification in the presence of water has been demonstrated under analytical conditions (Griffiths et al., 2010) and requires further optimization for large-scale use.

9.3.6 Impact of Species Selection

Richardson (2011) has demonstrated the importance of the selection of algal species on the environmental impact of the biodiesel process. Factors to consider include the specie's ability to scavenge light and CO_2, growth rate, lipid content, cell size, cell wall strength, digestibility, ability to settle without flocculation, as well as the nitrogen content of the biomass. The combination of these influences the volume of the process, the relative recovery of biodiesel to biogas, the selection of downstream processing equipment, and the need to pretreat the algae prior to fermentation or digestion.

9.3.7 The Biorefinery

During the production of algal biodiesel, an algal cake remains and can form a feedstock for further product formation. Potential commodity by-products include an algal biomass suitable for animal feed, fermentation of the carbohydrate portion of the algal biomass to ethanol (Sandler and Murthy, 2010), or anaerobic digestion of the algal cake to biogas. Here we consider the co-production of algal biogas to demonstrate the multi-product approach (Campbell et al., 2010; Stephenson et al., 2010; Richardson et al., 2012). Stephenson et al. (2010) investigated two modes of cell disruption and assumed that only disrupted cells were digested. While the anaerobic digestion of *Spirulina platensis* was similar in the presence and absence of cell disruption, disruption was essential for *Scenedesmus,* which also exhibited significant resistance to disruption. Collet et al. (2010) demonstrated anaerobic digestion of the algae to biogas without prior recovery of the lipid fraction for biodiesel. This approach allowed avoidance of the concentration and cell disruption steps, thereby reducing the energy and economic costs. The digestate following methane production has potential to provide a source of N and P for further algal growth. Partial recycle has been demonstrated in the *Spirulina* system.

9.4 ECONOMIC CONSIDERATIONS OF MICROALGAL PRODUCTION

9.4.1 COST OF ALGAL BIOMASS AND ALGAL OIL

Currently, commercial processes for the production of microalgae only exist for the production of specialty products such as health supplements, carotenoids, and specific aquaculture feeds. Furthermore, most of the algal species used in these processes are *extremophilic* algae. Data on the cost of large-scale algal processes are largely based on pre-implementation costing of these. Typical costs estimated are presented in Table 9.4.

From Table 9.4, the disparity in costing that arises from an immature technology position is clear. This was also noted in a comparison of environmental analyses. As a general trend, the algal biomass cost from the raceway system lies in the range of US\$0.23 to \$0.60 kg^{-1} DW, with the exception of the estimate of Norsker et al. (2011), for which the biomass was recovered by centrifugation, noted as a major capital and energy cost. The production of algal biomass from photobioreactors was characterized by a greater variation from US\$0.42 to \$3.04 kg^{-1} DW. Here it is evident that the two lowest values are based on the same calculations (Chisti, 2007), while the majority of the values (four of the eight available) lie in the range US\$3.18 to \$9.54. Refinement in this costing is required. Factors impacting the costing are discussed in Section 9.4.2.

These costs are not competitive with the cost of crude oil (US\$0.48 to \$0.71 L^{-1}; US\$76 to \$113 bbl^{-1}). Van Harmelen and Oonk (2006) estimated that cost of production with current technology exceeded potential earnings from algal oil as sole product by 1.7- to 3.9-fold, depending on the assumptions and process decisions made. In costing a production facility for algal oil in South Colorado, Richardson et al. (2012b) indicated that a 60% reduction in CAPEX (capital expenditure) and 90% reduction in OPEX (operating expenditure) are required for a cost-effective pond system with algal oil as the sole product. These reductions increase to 80% on CAPEX and 90% on OPEX using a photobioreactor system. Norsker et al. (2011) translated their costs based on algal biomass to an energy-based cost, yielding values in the range of US\$32.60 to \$295.50 GJ^{-1}. It is noted that the lower value of US\$32 GJ^{-1} estimated for a high light intensity environment is similar to the cost of delivered electricity.

These costings raise clearly two key considerations:

1. The necessity to operate the algal oil process as a biorefinery, achieving value from multiple products for the same growth costs
2. The need to interrogate component process costs to identify key targets for cost savings and technology development

TABLE 9.4

Cost Estimates for the Production of Algal Biomass and Algal Oil on a Large Scale

Ref.	Species	Location	Biomass Productivity (g DW m⁻²d⁻¹) [Biomass concentration (g L⁻¹)]	Lipid Content (%)	Estimated Cost of Algal Biomass (US$ per kg biomass)	Estimated Cost of Algal Lipid (L) or Biodiesel (D) (US$ per liter)	Assumptions wrt Process
Benemann and Oswald, 1996	–		30	50	0.241	0.43 (D)	Raceway
Benemann and Oswald, 1996	–		60	50	0.148	0.26 (D)	Raceway
Molina Grima et al., 2004	*Phaeodactylum*	–	–	–	30.4		Photobioreactor
Van Harmelen and Oonk, 2006			27	30	0.37	1.06 (D)	Raceway
Chisti, 2007	–		72	30	0.47	1.41 (D)	Photobioreactor
Chisti, 2007	–		35	30	0.60	1.81 (D)	Raceway
Norsker et al., 2011	*Spirulina, Dunaliella*	Eindhoven, the Netherlands	[0.3 g L⁻¹]		6.45		Raceway ponds, as wet paste. Paddlewheel circulating liquid at 0.25 m s⁻¹
Norsker et al., 2011	–	Eindhoven, the Netherlands	[2.1 g L⁻¹]		7.77		Flat-plate photobioreactor, as wet paste; aeration at 1 vvm; polyethylene film panels of 1-year lifetime

(Continued)

TABLE 9.4 (Continued)
Cost Estimates for the Production of Algal Biomass and Algal Oil on a Large Scale

Ref.	Species	Location	Biomass Productivity (g DW m⁻²d⁻¹) [Biomass concentration (g L⁻¹)]	Lipid Content (%)	Estimated Cost of Algal Biomass (US$ per kg biomass)	Estimated Cost of Algal Lipid (L) or Biodiesel (D) (US$ per liter)	Assumptions wrt Process
Norsker et al., 2011	*Haematococcus, Nannochloropsis, Chlorella*	Eindhoven, the Netherlands	[1.7 g L⁻¹]		5.41		Tubular photobioreactor, as wet paste; disposable polyethylene tubing; 1-year lifetime assumed; circulation at 0.5 m s⁻¹
Norsker et al., 2011	*Haematococcus, Nannochloropsis, Chlorella*	Bonaire, Dutch Antilles	[1.7 g L⁻¹]		3.18		Tubular photobioreactor, as wet paste; disposable polyethylene tubing; 1-year lifetime assumed; circulation at 0.5 m s⁻¹
Norsker et al., 2011	*Haematococcus, Nannochloropsis, Chlorella*	Bonaire, Dutch Antilles	[1.7 g L⁻¹]		0.91		Tubular photobioreactor, as wet paste; disposable polyethylene tubing; 1-year lifetime assumed; circulation at 0.5 m s⁻¹; assume free supply of CO₂ from flue gas, nutrients from wastewater, and photosynthetic efficiency increased to 60%

Reference	Species	Location						Notes
Jorquera et al., 2010	*Nannochloropsis*				0.227			Raceway system using Chisti (2008) and a lipid content of 29.6%
Jorquera et al., 2010	*Nannochloropsis*				0.419			Flat-plate reactor system using Chisti (2008) and a lipid content of 29.6%
Jorquera et al., 2010	*Nannochloropsis*				9.54			Horizontal tubular system using Chisti (2008) and a lipid content of 29.6%
Williams and Laurens, 2010		USA	18–37	15–50		0.36–0.65	0.79–3.08 (D)	Based on costings prepared for US Dept. of Energy's Office of Energy Efficiency and Renewable Energy using a hybrid reactor system consisting of a front-end PBR "nursery" and "grow-on" raceway stage
Richardson et al., 2012	—	South Colorado, USA					3.36 (L)	Open ponds
Richardson et al., 2012	—	South Colorado, USA					8.34 (L)	Photobioreactors

The effect of process variables and process selection on the costing is discussed in Section 9.4.2, while the potential contribution of products other than biodiesel is discussed in Section 9.4.3.

9.4.2 KEY PROCESS COMPONENTS CONTRIBUTING TO COST

It is well recognized that the production phase is most significantly affected by its energy requirements. This is most marked for photobioreactors and contributes significantly to costs. Norsker et al. (2011) explored these interactions. In the horizontal tubular reactor, liquid circulation contributes to both capital costs through the pump required and to energy costs. By reducing the linear velocity from 0.5 m s^{-1} to the minimum value predicted of 0.3 m s^{-1}, a reduction in cost per kilogram of 25% was achieved. Similarly, the aeration of the flat-plate reactor (or vertical tubular reactor) contributes significantly to capital and operating costs. In the former, a reduction in the aeration rate from 1 vvm (volume per volume per minute) to 0.3 vvm resulted in a 48% reduction in cost. The contributions of mixing and mass transfer in the flat-plate and tubular reactors were on the order of 52% and 30%, respectively.

Productivity significantly influences the cost of production. This is well demonstrated by the increased productivity with increasing light intensity, as correlated by Williams and Laurens (2010). Areal productivity increased from some 10 g m^{-2}d^{-1} at an irradiance of 15 MJ m^{-2}d^{-1} to 30 g m^{-2}d^{-1} on doubling the irradiance to 30 MJ m^{-2} d^{-1}. The impact of improved productivity under improved irradiance is illustrated by comparing the cost for production in Eindhoven (the Netherlands) and Bonaire (Dutch Antilles). In the three cases analyzed (Norsker et al., 2011), the cost decreased by 40% to 45% under conditions of increased illumination (summer high of 7,000 and 5,000 W-h m^{-2}d^{-1}, respectively; winter low of 4,500 and <1,000 W-h m^{-2}d^{-1}, respectively, in Bonaire and Eindhoven).

The provision of CO_2 can significantly impact the costing, based on whether the CO_2 is provided "free" as a by-product of an adjacent process or requiring its purchase as a compressed gas (Williams and Laurens, 2010). Stephenson et al. (2010) further considered the compression energy, with associated costs, based on the CO_2 concentration in the gas stream.

Williams and Laurens (2010) noted that their energy costs (typical of the US environment) were some sixfold higher than those estimated by an analysis conducted in British Columbia, Canada, where hydroelectric power was used. This highlights the potential for the use of renewable energy resources in conjunction with algal production.

9.4.3 POTENTIAL EARNINGS FROM BY-PRODUCTS

Potential by-products from the production of algal biodiesel include the capture of carbon, the generation of biogas through anaerobic digestion, the generation of bioethanol through fermentation of carbohydrates, the production of feeds for aquaculture, the production of animal feeds rich in protein and carbohydrates (~60% protein), and the production of a high-grade, protein-rich material (~90% protein). In the case of fermentation or digestion of the algal cell debris, the recycling of N and P media components can be used to decrease media costs.

Williams and Laurens (2010) considered the scenarios of biogas formation, preparation of animal feed, and preparation of high-grade, protein-rich material. They highlighted the uncertainty and process specificity of costing these options and noted the requirement for improved knowledge in this area. The 60% protein feed was valued at US$750 tonne^{-1} based on FAO (Food and Agriculture Organization) statistics. Relating it to the value of soya meal containing 45% protein, a value of US$500 tonne^{-1} was proposed. In comparison, the high-grade, protein-rich material was valued at US$900 tonne^{-1}. Biodiesel was estimated at a value of US$125 bbl^{-1}. Using this estimate, positive scenarios were found at protein feed values of US$350 tonne^{-1} and above, and protein-rich extracts at US$600 tonne^{-1} and above. Under the conditions used, anaerobic digestion was not cost effective; however, the high energy requirement to maintain the digester temperature was a result of the temperate environment, and revised analysis is required for warmer climates where the performance of biodigesters is well documented.

9.5 KEY FOCAL AREAS FOR IMPROVING ENVIRONMENTAL AND ECONOMIC SUSTAINABILITY

The motivation to overcome the challenges with respect to environmental and economic sustainability of microalgal culture for biodiesel production and other renewable products stems from the significant advantages of the microalgal system as a biomass source. These include the potential to use nonarable land for microalgal cultivation, the homogeneity of the biomass formed and the ability to process all components, the much-improved oil production per unit area (15 to 300 times greater), the higher growth rate, and the photosynthetic efficiency of microalgae compared with terrestrial plants (up to tenfold increase) (Chisti, 2007; Schenk, 2008; Rodolfi et al., 2009). Freshwater, seawater, brines, and wastewater are all potential water sources for algal growth (Vasudevan and Briggs, 2008). CO_2 uptake by the autotrophic algae enables CO_2 cycling through uptake for biomass generation and release on fuel combustion. Furthermore, the multiple energy forms attainable from microalgae span liquid fuels, and heat and electricity generation, enabling ongoing support of existing technologies while developing a reduced carbon economy. On attaining an energy economy in which dependence on carbon combustion is reduced, the technology lessons learned through achieving environmental and economically sustainable algal biomass will readily be transferred to the production of carbon-based commodities with simultaneous carbon sequestration or cycling.

The NER and LCA studies conducted to date have highlighted the great sensitivity of the GWP, fossil fuel requirements, and NER on the productivity of algal biomass and of algal oil attainable in the algal cultivation process. While maximum specific growth rates and lipid content are partly defined through the algal species selected, culture conditions may be used to enhance these through improved light supply, mass transfer, and mixing. The energy requirement of the bioreactor to achieve mixing and mass transfer is a major contributor to the energy requirement of the integrated algal process, as is the CO_2 provision to the reactor. Based on the volumetric concentrations attainable, pumping energies can be defined.

Supply of nutrients, yield of products from these nutrients, and recycle of unused nutrients impact GWP and fossil fuel requirements significantly. Opportunities exist in selecting algal species able to scavenge low nutrient concentrations, having reduced nitrogen content, as well as to utilize nitrogen and phosphorous resources efficiently, either by their provision from wastewater or through their recycle.

The typically dilute biomass concentrations required to minimize light limitation result in the need to process large culture volumes; hence, natural flocculation to facilitate settling is beneficial. In downstream processing, the most important factors pertain to the ability to process wet biomass, thereby eliminating the drying process, as well as the ability to recover product from the algal cell readily, thus minimizing the requirement for conventional, energy-intensive cell disruption.

As reported by Harding (2009), the production phase typically has the greatest impact on the overall LCA; hence, its optimization is required in the first instance.

Opportunities to improve the economics correlate well with the environmental analysis with respect to operating costs. These highlight mass transfer and mixing, provision of CO_2, provision of nutrients, biomass recovery, and the avoidance of rigorous drying. More importantly, the economic studies suggest that algal biofuel technology requires further enhancement prior to its economic feasibility as a stand-alone technology. However, opportunity exists to establish a cost-effective algal biorefinery delivering a combination of products such as biodiesel, biogas, animal feed, and protein extracts. To this, high-value products may be added. Further, the predicted cost of algal biomass positions it attractively as a raw material source for bulk products, including fuels, chemicals, materials feeds, and food supplements.

REFERENCES

Batan, L., Quinn, J., Willson, B., and Bradley, T. (2010). Net energy and greenhouse gas emission evaluation of biodiesel derived from microalgae. *Environmental Science and Technology,* 44(20): 7975–7980.

Benemann, J., and Oswald, W.J. (1996). *Systems and Economic Analysis of Microalgae Ponds for Conversion of CO$_2$ to Biomass.* Pittsburgh Energy Technology Center: U.S. DOE.

Campbell, P.K., Beer, T., and Batten, D. (2010). Life cycle assessment of biodiesel production from microalgae in ponds. *Bioresource Technology,* 102(1): 50–56.

Chisti, Y. (2007). Biodiesel from microalgae. *Biotechnology Advances,* 25(3): 294–306.

Chisti,Y., and Yan, J. (2011). Energy from algae: Current status and future trends. Algal biofuels – A status report. *Applied Energy* 88: 3277–3279.

Clarens, A.F., Resurreccion, E.P., White, M.A., and Colosi, L.A. (2010). Environmental life cycle comparison of algae to other bioenergy feedstocks. *Environmental Science and Technology,* 44(5): 1813.

Collet, P., Hélias, A., Lardon, L., Ras, M., Goy, R., and Steyer, J. (2010). Life-cycle assessment of microalgae culture coupled to biogas production. *Bioresource Technology,* 102(1): 207–214.

Consoli, F., Allen, D., Boustead, I., Fava, J., Franklin, W., Jense, A.A., De Oude, N., Parrish, R., Perriman, R., Postlewaite, D., Quay, B., Sequin, J., and Vigon, B. (Eds.) (1993). *Guidelines for Life-Cycle Assessment: A Code of Practice.* SETAC Publications, Pensacola, FL.

Curran, M.A. (2000). Life cycle assessment: An international experience. *Environmental Progress,* 19(2): 65–71.

Evans, A., Strezov, V., and Evans, T.J. (2009). Assessment of sustainability indicators for renewable energy technologies. *Renewable and Sustainable Energy Reviews,* 13(5): 1082–1088.

Fargione, (2008). Land clearing and the biofuel carbon debt. *Science,* 319(5867): 1235.

Griffiths, M.J., and Harrison, S.T.L. (2009). Lipid productivity as a key characteristic for choosing algal species for biodiesel production. *Journal of Applied Phycology*, 21(5): 493–507.

Griffiths, M.J., Van Hille, R.P., and Harrison, S.T.L. (2010). Selection of direct transesterification as the preferred method for assay of fatty acid content of microalgae. *Lipids,* 45: 1053–1060.

Harding, K.G. (2009). A Generic Approach to Environmental Assessment of Microbial Bioprocesses through Life Cycle Assessment. Ph.D. thesis, Department of Chemical Engineering University of Cape Town, Cape Town, South Africa.

Harding, K.G., Dennis, J.S., and Harrison, S.T.L. (2012). Material and energy balance and life cycle assessment study on Penicillin V production using a generic flowsheet model approach for first estimate studies. *Journal of Biotechnology* (submitted).

Harding, K.G., Dennis, J.S., Von Blottnitz, H., and Harrison, S.T.L. (2008). A life-cycle comparison between inorganic and biological catalysis for the production of biodiesel. *Journal of Cleaner Production,* 16(13): 1368–1378.

International Association for Impact Assessment. (1999). Principle of Environmental Impact Assessment Best Practice. IAIA, Fargo, ND.

ISO 14040: 2006 (2006). Environmental Management – Life Cycle Assessment – Principles and Framework. International Organization for Standardization, Geneva, Switzerland.

Jorquera, O., Kiperstok, A., Sales, E.A., Embiruçu, M., and Ghirardi, M.L. (2010). Comparative energy life-cycle analyses of microalgal biomass production in open ponds and photobioreactors. *Bioresource Technology,* 101(4): 1406–1413.

Kadam, K.L. (2002). Environmental implications of power generation via coal-microalgae cofiring. *Energy.* 27(10): 905–922.

Kaltschmitt, M., Reinhardt, G.A., and Stelzer, T. (1997). Life cycle analysis of biofuels under different environmental aspects. *Biomass and Bioenergy,* 12(2): 121–134.

Kaparaju, P., Serrano, M., Thomsen, A.B., Kongjan, P., and Angelidaki, I. (2009). Bioethanol, biohydrogen and biogas production from wheat straw in a biorefinery concept. *Bioresource Technology,* 100(9): 2562–2568.

Kim, S., and Dale, B.E. (2005). Life cycle assessment of various cropping systems utilized for producing biofuels: Bioethanol and biodiesel. *Biomass and Bioenergy,* 29(6): 426–439.

Lardon, L., Helias, A., Sialve, B., Steyer, J., and Bernard, O. (2009). Life-cycle assessment of biodiesel production from microalgae. *Environmental Science & Technology,* 43(17): 6475–6481.

Molina Grima, E., Garcia Camacho, F., Sanchez Perez, J.A., Fernandez Sevilla, J.M., Ancien Fernandez, F.G., and Contreras Gomez, A. (2004). A mathematical model of micro-algal growth in light-limited chemostat culture. *Journal of Chemical Technology and Biotechnology*, 61(2): 167–173.

Norsker, N.-H, Barbosa, M.J., Vermue, M.H., and Wijffels, R.H. (2011). Microalgal production – A close look at the economics. *Biotechnology Advances,* 29: 24–27.

Razon, L.F., and Tan, R.R. (2011). Net energy analysis of the production of biodiesel and biogas from the microalgae: *Haematococcus pluvialis* and *Nannochloropsis. Applied Energy,* 88: 3507–3514.

Richardson, C. (2011). Investigating the Role of Reactor Design to Maximize the Environmental Benefit of Algal Oil for Biodiesel. M.Sc. dissertation, Department of Chemical Engineering, University of Cape Town, South Africa.

Richardson, C., Griffiths M.J., Von Blottnitz, H., and Harrison S.T.L. (2012a). Investigating the role of reactor design for maximum environmental benefit of algal oil for biodiesel. *Bioresource Technology* (submitted).

Richardson, J.W., Johnson M.D., and Outlaw J.L. (2012b). Economic comparison of open pond raceways to photo bio-reactors for profitable production of algae for transportation fuels in the Southwest. *Algal Research* 1(1): 95–100.

Rodolfi, L., Zittelli, G.C., Bassi, N., Padovani, G., Biondi, N., Bonini, G., and Tredici, M.R. (2009). Microalgae for oil: Strain selection, induction of lipid synthesis and outdoor mass cultivation in a low-cost photobioreactor. *Biotechnology and Bioengineering,* 102(1): 100–112.

Sander, K., and Murthy, G.S. (2010). Life cycle analysis of algae biodiesel. *International Journal Life Cycle Assessment,* 15: 704–714.

Sazdanoff, N. (2006). Modeling and Simulation of the Algae to Biodiesel Fuel Cycle. Undergraduate thesis. United States: Department of Mechanical Engineering, The Ohio State University.

Schenk, P.M. (2008). Second generation biofuels: High-efficiency microalgae for biodiesel production. *Bioenergy Research,* 1(1): 20–43.

Searchinger, (2008). Use of US croplands for biofuels increases greenhouse gases through emissions from land-use change. *Science,* 319(5867): 1238–1240.

Sheehan, J., Dunahay, T., Benemann, J., and Roessler, P. (1998). A Look Back at the U.S. Department of Energy's Aquatic Species Program: Biodiesel from Algae. Close-Out Report. Golden, CO: National Renewable Energy Lab, Department of Energy.

Stephenson, A.L., Kazamia, E., Dennis, J.S., Howe, C.J., Scott, S.A., and Smith, A.G. (2010). Life-cycle assessment of potential algal biodiesel production in the United Kingdom: A comparison of raceways and air-lift tubular bioreactors. *Energy and Fuels,* 43(17): 4062–4077.

UK Carbon Trust (2008). Carbon Footprinting. Carbon Trust, London, UK.

Van Harmelen, T., and Oonk, H. (2006). Micro-algae Bio-fixation Processes: Applications and Potential Contributions to Greenhouse Gas Mitigation Options. Report for the International Network on Bio-fixation of CO_2 and Greenhouse Gas Abatement with Micro-algae operated under the International Agency Greenhouse Gas R&D Programme.

Vasudevan, P., and Briggs, M. (2008). Biodiesel production-current state of the art and challenges. *Journal of Industrial Microbiology Biotechnology,* 35(5): 421–430.

Von Blottnitz H., and Curran M.A. (2007). A review of assessments conducted on bio-ethanol as a transportation fuel from a net energy, greenhouse gas, and environmental life cycle perspective. *Journal of Cleaner Production,* 15: 607–619.

Wackernagel, M., Schulz, N.B., Deumling, D., Linares, A.C., Jenkins, M., Kapos, V., Monfreda, C., and Loh, J. (2002). Tracking the Ecological Overshoot of the Human Economy. *Proceedings of the National Academy of Sciences of the United States of America,* 99(14): 9266–9271. doi: 10.1073/pnas.142033699.

Williams, P.J.le B., and Laurens, L.M.L. (2010). Microalgae as biodiesel and biomass feedstocks: Review and analysis of the biochemistry, energetics and economics. *Energy Environmental Sciences,* 3: 554–590.

Yang, J., Xu, M., Zhang, X., Hu, Q., Sommerfeld, M., and Chen, Y. (2011). Life-cycle analysis on biodiesel production from microalgae: Water footprint and nutrients balance. *Bioresource Technology,* 102(1): 159–165.

10 Value-Added Products from Microalgae

Terisha Naidoo, Nodumo Zulu,
Dheepak Maharajh, and Rajesh Lalloo
CSIR Biosciences
Pretoria, South Africa

CONTENTS

10.1 INTRODUCTION

Microalgae represent a biodiverse resource (Metting, 1996; Pulz and Gross, 2004). The complexity of their chemical composition and range of biochemical products make these organisms exploitable resources for valuable and novel products in the food, feed, pharmaceutical, and research industries (Pulz and Gross, 2004). The market for these applications is still emerging, but there have already been new areas of research in microalgal biotechnology to satisfy the new product demands of industry and consumers.

The use of microalgae as food for human consumption is an age-old tradition. Over 2,000 years ago, the Chinese used *Nostoc* to survive famine (Milledge, 2011). Species such as *Arthrospira* (*Spirulina*) and *Aphanizomenon* have also been utilized for decades as a source of food (Spolaore et al., 2006). Despite the plethora of historical usage by humans, microalgal culture is a fairly new area of biotechnology research, and its commercial application is virtually untapped (Spolaore et al., 2006; Milledge, 2011).

The interest in algal biomass came about in the 1950s as a result of an increase in the world's population, and a forecast of insufficient protein supply triggered the search for alternative novel protein sources (Spolaore et al., 2006). Today, the use of algae for food still continues in many parts of the world; however, the large-scale production of algae to eradicate the food calorie and protein shortage has not fully materialized (Milledge, 2011).

There are a limited number of revolutionary companies that have persevered to large-scale production of algal biomass and products. The algal products are normally marketed by the dominant players in the food and pharmaceutical industries. However, there is a significant gap in the microalgal market for expansion of existing products and the introduction of new products (Luiten et al., 2003; Pulz and Gross, 2004; Becker, 2007).

10.2 COMMERCIALLY EXPLOITED MICROALGAE, PRODUCTS, AND APPLICATIONS

At present, the most significant product of microalgal biotechnology in terms of production amounts and economic value is microalgal biomass (Figure 10.1; see color insert). Microalgal biomass has been widely used in the fuel and energy sectors; however, the nutritional value of algal biomass has endorsed its use as a high-protein supplement in human nutrition, aquaculture, and as a nutraceutical (Del Campo et al., 2007).

The algal biomass market size is estimated at around 10,000 tonnes y^{-1} (dry weight) (Becker, 2007), with an annual turnover of over US\$1.25 million (Milledge, 2011). Despite being a biodiverse resource, microalgae remain understudied in terms of their morphology and physiology. Much of the literature regarding the systematics and taxonomy of microalgae focuses on biotechnologically relevant species. More than 200,000 species are known to exist; however, only 10 to 20 species (Table 10.1) have been exploited worldwide for biomass, pigments, antioxidants, and special products (toxins and isotopes) for various product applications (Borowitzka, 1992; Radmer, 1996; Olaizola, 2003).

This chapter addresses in detail various types of high-value products derived from algal biomass, their respective applications, production systems, and market positions.

10.2.1 CAROTENOIDS

Carotenoids are colored, lipid-soluble compounds that occur in higher plants, microalgae, as well as in nonphotosynthetic organisms (Del Campo, 2007; Takaichi, 2011). Carotenoids contribute to light harvesting, maintenance of structure, and functioning

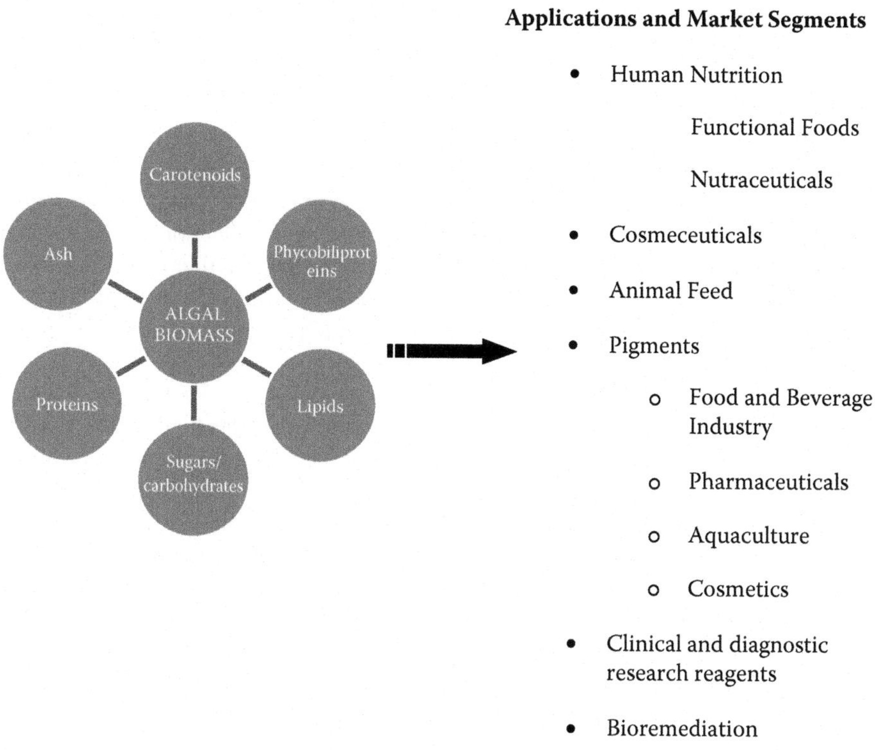
Carotenoids
Ash
Phycobiliprot eins
ALGAL BIOMASS
Proteins
Lipids
Sugars/ carbohydrates
Applications and Market Segments
Human Nutrition
Functional Foods
Nutraceuticals
Cosmeceuticals
Animal Feed
Pigments
Food and Beverage Industry
Pharmaceuticals
Aquaculture
Cosmetics
Clinical and diagnostic research reagents
Bioremediation

FIGURE 10.1 (See color insert.) Applications of algal biomass.

TABLE 10.1
Global Production of Algal Biomass for Commercially Relevant Algal Genera

Genus	Production (tonnes y^{-1} dry weight)	Country	Applications and Products
Spirulina	3,000	China, India, U.S., Myanmar, Japan, SA	Human and animal nutrition, phycobiliproteins, cosmetics
Chlorella	2,000	Taiwan, Germany, Japan	Human nutrition, aquaculture, cosmetics
Dunaliella	1,200	Australia, Israel, U.S., China, SA	Human nutrition, cosmetics, β-carotene
Nostoc	600	China	Human nutrition
Aphanizomenon	500	U.S.	Human nutrition
Haematococcus	300	U.S., India, Israel	Aquaculture, astaxanthin

Source: Adapted from Pulz and Gross (2004), Spolaore et al. (2006); Milledge (2011).

of photosynthetic complexes in plants and microalgae (Pulz and Gross, 2004; Del Campo, 2007). They occur widely in nature and are responsible for many of the brilliant red, orange, and yellow colors of edible vegetables and fruits and some aquaculture animals.

Microalgae combine properties of higher plants with some properties of prokaryotes. This combination represents the rationale for using microalgae for the production of carotenoids and other products (Del Campo, 2007; Guedes et al., 2011) instead of using plants or prokaryotes. Furthermore, the production of carotenoids by microalgae can be easily maximized by manipulating growth conditions. Under unfavorable growth conditions, microalgae produce high amounts of carotenoids, such as β-carotene, astaxanthin, and canthaxanthin (Orosa et al., 2000).

10.2.1.1 Commercial Applications

There is a demand for natural pigments to be applied in the food, pharmaceutical, and aquaculture industries (Dufossé et al., 2005). The use of synthetic dyes in these industries is slowly declining due to their toxic effects (Dufossé et al., 2005). Compared to synthetic alternatives, microalgal carotenoids have the advantage of supplying natural isomers in their natural ratios (Pulz and Gross, 2004; Milledge, 2011). Microalgal pigments have been used as alternatives to synthetic pigments in various industries (Table 10.2).

Other than for coloring purposes, carotenoids have recently been used as antioxidants. Carotenoids have antioxidant effects that can be beneficial in countering diseases such as cancer, obesity, and hypertension (Inbaraj et al., 2006; Murthy et al., 2005). Table 10.3 indicates the applications of commercially exploited carotenoids in various industries.

The worldwide demand for carotenoids has been increasing at an average yearly rate of 2.2% (Guedes et al., 2011). Among the 400 known carotenoids, so far only a few have been exploited (Cosgrove, 2010; Milledge, 2011). The two most commonly exploited algal carotenoids are β-carotene and astaxanthin, which are mainly produced by *Dunaliella salina* and *Haematococcus pluvialis,* respectively (Pulz and Gross, 2004; Spolaore et al., 2006; Del Campo, 2007; Cosgrove 2010;

TABLE 10.2
Utilization of Microalgae for Production of Natural Pigments

Microalgal Strain	Production System	Products Used as Natural Pigments	Ref.
Spirulina	Open pond	Phycocyanin	Dufossé et al., 2005
Dunaliella salina	Open pond	β-Carotene	Dufossé et al., 2005; Del Campo et al., 2007
Haematococcus pluvialis	Open pond	Astaxanthin	Dufossé et al., 2005; Del Campo et al., 2007
Muriellopsis sp.	Open pond	Lutein	Dufossé et al., 2005; Del Campo et al., 2007

TABLE 10.3

Commercially Available Carotenoids and Their Applications

Carotenoid Type	Food Coloring	Coloring of Animals and of Animal Products	Supplements (anti-oxidants)	Vitamin A Source	Cosmetic	Ref.
Annatto	X					Vital Solutions market report, 2010
Apocarotenal	X					Vital Solutions market report, 2010
Apocarotenal ester		X				Vital Solutions market report, 2010
Astaxanthin	X	X	X		X	Pulz and Gross, 2004; Vital Solutions market report, 2010; Milledge, 2011
Beta-carotene	X	X	X	X	X	Pulz and Gross, 2004; Spolaore et al., 2006; Del Campo, 2007; Vital Solutions market report, 2010; Milledge, 2011
Canthaxanthin		X				Inbaraj et al., 2006; Vital Solutions market report, 2010
Capsanthin	X	X	X			Vital Solutions market report, 2010
Lutein	X	X	X			Del Campo, 2007; Vital Solutions market report 2010
Lycopene			X		X	Vital solutions market report, 2010
Zeaxanthin	X		X			Del Campo, 2007; Vital Solutions market report 2010

Milledge, 2011). The market size for β-carotene is estimated at 1,200 tonnes per year and greater than US$280 million in sales volume per year (Pulz and Gross, 2004). The market price of natural β-carotene is much higher than that of synthetic β-carotene ($1,000 to $2,000 kg^{-1} for natural β-carotene versus $400 to $800 kg^{-1} for synthetic β-carotene). Although the price of natural β-carotene is higher than that of the synthetic form, preference is still given to the natural form because it has physical properties that make it superior to the synthetic form. The annual worldwide market of astaxanthin is estimated at US$200 million (Spolaore et al., 2006). The market size of astaxanthin is estimated at just below 300 tonnes per year, with a sales volume of less than US$150 million per year (Pulz and Gross, 2004; Spolaore et al., 2006).

Lutein and zeaxanthin are the other two carotenoids with great potential. These carotenoids are commonly derived from petals of *Tagetes erecta* and *Tagetes patula*, commonly referred to as marigold flowers (Del Campo et al., 2007). Microalgae also have an ability to accumulate these carotenoids. Lutein and zeaxanthin are known to selectively accumulate in the macula of the human retina. They protect the eyes from light and oxidative stresses (Kotake-Nara and Nagao, 2011).

The lutein extracted from other sources is usually 95% esterified, whereas in microalgae, lutein is found in the free nonesterified form. *Muriellopsis* sp. recorded lutein yields of 75 mg m^{-2}d^{-1} in an outdoor open pond system (Del Campo et al., 2007). These values are similar to those obtained in a closed system (Harun et al., 2010; Del Campo et al., 2007). Overall, the free lutein content of *Muriellopsis* sp. biomass varies between 0.4% and 0.6%; which represents a higher content of esterified lutein than found in crown petals of *Tagetes* plants. The global market size of lutein is expected to hit $124.5 million by 2013 (Heller, 2008).

Zeaxanthin is mainly produced synthetically due to the fact that its content in natural sources (including microalgae) is considered very low for industrial production. The major problem underlying the commercial exploitation of zeaxanthin is the development of production processes that will result in the extraction of high amounts of zeaxanthin (Weiss et al., 2008). In 2006, the global market of zeaxanthin was estimated at $2 million (Heller, 2008). The awareness of this carotenoid still remains lower than that of lutein.

10.2.1.2 Production Processes

Carotenoid production usually occurs in open pond raceway systems and in photobioreactors, depending on the robustness of the algal strain toward contamination and the purity requirements or application of the final product. Carotenoids such as astaxanthin, lutein, and zeaxanthin are produced in photobioreactors (Dufossé et al., 2005; Milledge, 2011) due to the sensitivity of the algal strains to contamination. Algal strains that have an ability to grow in harsh environments are usually cultivated in open pond raceway systems. As an example, *Dunaliella salina*, for the production of β-carotene, can grow in high-salinity environments (Dufossé et al., 2005).

Biomass harvesting methods also depend on the algal strain cultivated. The preferred methods of harvesting biomass for carotenoid production are centrifugation, sedimentation, and filtration (Dufossé et al., 2005; Weiss et al., 2008). Subsequent to centrifugation, the hard cell walls are broken and then extraction of the carotenoid occurs (Dufossé et al., 2005).

10.2.1.3   Foresight

Carotenoid production has established itself as the most successful area of microalgal biotechnology; and with the increasing market demands for these natural pigments, the future of microalgal carotenoid production appears promising (Del Campo et al., 2007). The ability of microalgae to be genetically modified opens doors for enhancing specific carotenoid production through metabolic engineering. However, this approach might not be welcomed by the food and aquaculture industries due to the controversy surrounding genetically modified products. The market demand for carotenoids is expected to increase even further with the discovery that carotenoids exhibit tumor-suppressing activity (Schmidt-Dannert et al., 2000). Carotenoid exploitation is restricted to only a few algal species; more algal strains have yet to be screened.

10.2.2   PHYCOBILIPROTEINS

Phycobiliproteins are photosynthetic accessory pigments produced by microalgae. These pigments are responsible for improving the efficiency of light energy utilization (Pulz and Gross, 2004). Phycobiliproteins are deeply colored (red or blue), water-soluble complex proteins and have a broad spectrum of potential applications as natural coloring agents in the food and feed, pharmaceutical, and cosmetics industries. Among the cyanobacteria and red algae, there are four main classes of phycobiliproteins that are synthesized (Table 10.4): allophycocyanin (APC, bluish-green), phycocyanin (PC, blue), phycoerythrin (PE, purple), and phycoerythrocyanin (PEC, orange).

TABLE 10.4
Phycobiliprotein Content in Various Algal Strains

Species	Nature of Pigment	% Yield (dry weight)
Cyanobacteria		
Anabaena sp.	Phycocyanin	8.3
Nostoc sp.	Phycocyanin	20
Phormidium valderianum	Phycocyanin	20
Spirulina fusiformis	C-Phycocyanin	46
Spirulina platensis (syn *A. platensis*)	C-Phycocyanin	9.6
	Allophycocyanin	9.5
Red Algae		
Rhodosorus marinus	Phycoerythrin	8
Porphyridium cruentum	B-Phycoerythrin	32.7
	R-Phycocyanin	11.9

These phycobiliproteins are geometrically incorporated into structures called phycobilisomes, which are located on the outer surface of the thylakoid membranes.

10.2.2.1 Commercial Applications of Phycobiliproteins

The two most commercially exploited genera (Table 10.4) are the cyanobacterium *Arthrospira* sp. (*Spirulina* sp.) and the rhodophyte *Porphyridium* sp., which are responsible for the production of phycocyanin and phycoerythrin, respectively (Spolaore et al., 2006; Sekar and Chandramohan, 2008). Other commercially produced genera include *Rhodella* sp. and *Spirulina fusiformis*.

Powerful spectral properties make them suitable for use as highly sensitive fluorescent reagents in clinical or research immunology laboratories (Pulz and Gross, 2004; Spolaore et al., 2006; Sekar and Chandramohan, 2008; U.S. DOE, 2010; and Milledge, 2011). They also function as labels for antibodies and receptors among other biological molecules in a fluorescence-activated cell sorter and are used in immunolabeling experiments and fluorescence microscopy and diagnostics (Spolaore et al., 2006; Sekar and Chandramohan, 2008). A number of multinational companies (Table 10.5) have been contributing to the algal phycobiliprotein market, which is targeted at the medical and biotechnology research industry (Eriksen, 2008; Sekar and Chandramohan, 2008).

A Japan-based company, Dainippon Ink and Chemicals, is responsible for developing a product coined "Lina Blue," which is used extensively in the food industry (in chewing gum, ice slush, popsicles, candies, soft drinks, dairy products, and wasabi). A derivative of this pigment is also sold as a colorant for cosmetics such as

---

TABLE 10.5

Phycobiliprotein Products in the Commercial Sector and Medical and Biotechnology Research

Properties Reported in Patents	Commercial Distributors
Phycobiliproteins as fluorescent labels, tags, and markers	Cyanotech Corporation
	PROzyme Inc.
	Pierce Biotechnology Inc.
	Dojindo Molecular Technologies
	Flogen®
	ANAspec Inc.
	Martek Bioscience Corporation
	Invitrogen-Molecular Probes
Phycobiliprotein conjugates	Vector Laboratories
	Martek Bioscience Corporation
	Invitrogen-Molecular Probes
	Europa Bioproducts
Stabilized phycobilisome	Martek Bioscience Corporation

Source: Adapted from Sekar and Chandramohan (2008).

---

eyeliner and lipstick (Spoalore et al., 2006; Milledge, 2011). Although not produced commercially, the red alga *Porphyridium aerugineum* was used to produce a blue color that is added to Pepsi® and Bacardi Breezer® (Dufossé et al., 2005). It is no surprise that the global market for phycobiliprotein colorants alone was estimated at US$50 million by 2010 (Del Campo et al., 2007), with prices varying from US$3 to US$25 mg^{-1} (Spolaore et al., 2006; Milledge, 2011).

10.2.2.2 Production Process

Phycocyanin is employed as a colorant to a greater degree compared to phycoerythrin, which is incorporated more frequently in fluorescent applications. This is evident in their production yields (Table 10.4), where C-phycocyanin yields are reported to be as high as 46%, which is consistent with its broad application profile. C-phycocyanin (PC) is the source of blue coloring and is commercially produced from *Spirulina, Porphyridium,* and *Rhodella* (Milledge, 2011).

The majority of the commercial production of PC occurs in outdoor, photoautotrophic open raceway ponds predominantly in subtropical locations around the Pacific Ocean, specifically with *Spirulina platensis* (Spolaore et al., 2006; Eriksen, 2008). The range of commercial applications drives the production of high-purity phycobiliproteins—through extraction from the phycobilisomes followed by purification. The extraction process is particularly difficult because of the rigid cellular wall and the small size of the cell. Therefore, physical or chemical cell disruption is necessary to increase the bioavailability and assimilation of phycobiliproteins from the cells (Molina-Grima, 2003; Sekar and Chandramohan, 2008). There are a number of extraction methods available to aid in the cell disruption process, of which include sonication with sand (mainly small-particle silica), French press, tissue grinding (with or without liquid nitrogen), homogenization, and causing osmotic shock with use of dilute phosphate buffer. Upon comparing all the extraction methods tested, freezing and thawing of cells with liquid nitrogen, followed by grinding with a mortar and pestle (with an abrasive material) and homogenization at 10,000 rpm yielded almost 20% phycocyanin from *Spirulina* dry biomass (Sekar and Chandramohan, 2008).

There exists a range of patents detailing various cultivation and harvesting systems, extraction methods, and purification and production processes for phycobiliproteins. Purification of phycoerythrin includes distilled water leaching, staged precipitation with ammonium sulfate, and ion-exchange chromatography (Sekar and Chandramohan, 2008). Good-quality algal pigments, specifically with respect to color tone and thermal stability, were patented for use as colorants in food. Such pigments were obtained by evaporating an aqueous solution containing trehalose and algal pigments to dryness (Sekar and Chandramohan, 2008). Consistently and efficiently cultivating large amounts of algae throughout the year without being affected by conditions of the culturing site can be challenging. Thus, methods have been patented to proliferate the growth of algae by irradiating the culture with monochromatic light at a wavelength of 600 nm. Cultivation of cyanobacteria under a magnetic field for the production of phycobiliproteins was patented for *Spirulina* and *Colarina*. This involves charging the algae in a test tube, by placing the test tube between the N- and S-poles of a magnet, such that both poles oppose each other on both sides of the tube. For the production of phycobiliproteins, this is done under constant irradiation with a fluorescent lamp with an illuminance of 800 to 8,000 lux at 24°C for 480 h (Sekar and Chandramohan, 2008).

The utilization of urea-type or amino-type water-soluble nitrogen compounds, together with other required nutrients, has also been patented as a cultivation method to increase phycocyanin yields (Sekar and Chandramohan, 2008).

10.2.2.3   Future Potential

The broad application profile in the food industry and the increasing interest in fluorescent products showcase the diverse and promising potential of phycobiliproteins in number of applications. Table 10.5 exhibits several other novel properties of phycobiliproteins that have the potential for commercialization, but have only been accounted for in patents. Other biomedical properties included are anti-inflammatory, antioxidant, liver protection, anti-tumor, lipase activity inhibitor, and serum reducing agent; all of which have been reported in patents and applied research but have not yet been commercially explored (Sekar and Chandramohan, 2008).

Despite extensive research spanning 150 years, and the thousands of microalgal species that are known to exist, only a few hundred have been screened for chemical compositions and only a handful have been exploited on an industrial scale (Spolaore et al., 2006; Sekar and Chandramohan, 2008). To make phycobiliproteins more market competitive and economically feasible, basic screening is imperative in order to source the organisms that are responsible for significant production of phycobiliproteins—but may not necessarily be the fastest-growing strains (Sekar and Chandramohan, 2008). Genetic modification of microalgae holds great promise, along with pursuing other methods of cultivation (heterotrophic and mixotrophic). With the ever-increasing range of potential products and applications pending commercialization, it is imperative to pursue these avenues of research to advance microalgal biotechnology.

10.2.3   LIPIDS

Microalgae are responsible for the production of a range of lipids, with contents varying from 1% to 70% of dry weight, and reaching up to 90% (Metting, 1996, Spolaore et al., 2006). However, the most significant contribution to the overall microalgal market by algae is their ability to synthesize PUFAs (polyunsaturated fatty acids). Potential applications of various microalgal PUFAs are given in Table 10.6.

The omega-3 fatty acids eicosapentaenoic acid (EPA) and docosahexaenoic acid (DHA) are of particular interest as they cannot be efficiently synthesized by humans, and instead must be consumed in their diet (Simopoulos, 1999). Over the years and even today, omega-3 EPA and DHA are regarded as common constituents of fish oil. Table 10.6 presents microalgal producers of interesting PUFAs, and presently DHA is the only algal PUFA that is commercially available (Spolaore et al., 2006).

TABLE 10.6
Essential Microalgal PUFAs

PUFA	Potential Application	Microalgal Producer
γ-Linoleic acid (GLA) ω6	Infant formula (full-term infants) Nutraceuticals	*Arthrospira* (*Spirulina*) sp.
Arachidonic acid (AA) ω6	Infant formula (full-/pre-term infants) Nutraceuticals	*Porphyridium* sp.
Eicosapentaenoic acid ω3 (EPA)	Neutraceuticals Aquaculture	*Nannochloropsis* *Phaeodactylum* *Nitzschia* sp.
Docosahexaenoic acid ω3 (DHA).	Infant formula (full-/pre-term infants) Nutraceuticals Aquaculture	*Crypthecodinium* (*Schizochytrium*) sp.

10.2.3.1 Eicosapentaenoic Acid (EPA) and Docosahexaenoic Acid (DHA)

The two most significant essential fatty acids found in considerable levels in meat and coldwater fish are the omega-3 eicosapentaenoic acid (EPA) and docosahexaenoic acid (DHA) (see Table 10.7). New evidence suggests beneficial effects of omega-3 on diseases such as cardiovascular disease (CVD), inflammatory disease, and brain function. Recent studies have also shown the positive impact of omega-3 in curing mental health disorders (Simopoulos, 1999; Arterburn et al., 2000; Nemets et al., 2002; Kris-Etherton et al., 2003, Wen and Chen, 2003; Ruxton et al., 2004; Freeman et al., 2006; Von Schacky and Harris, 2007; Mischoulon et al., 2008; The Ocean Nutrition Canada website, 2010). Aside from human health, omega-3 has significant advantages for growth and development (Ruxton et al., 2004). As a result, more than 14,000 studies have been conducted over the past 35 years, promoting the benefits of omega-3 fatty acids in the human diet at every stage of life (The Ocean Nutrition Canada website, 2010).

Currently, algal EPA and DHA are the only alternative to fish oils. Apart from being a complete vegetarian alternative, microalgae are considered sustainable feedstocks for the production of EPA and DHA compared to other sources such as sardines, krill, and genetically engineered oilseed crops. Consumers are well aware of the choices they make, and they prefer an omega-3 source that is "naturally biodiverse and not genetically engineered" (Watson, 2011a). In general, algal cultures are pure and eliminate concerns about high levels of toxins, pollutants, and heavy metals. Algal oils have a high unsaturation index and with the aid of novel processing techniques to improve stabilization by reducing oxidation potential, algal oils provide fish-free odor and taste. New and improved technologies, such as microencapsulation, allow these oils to be employed in a broader food and beverage application profile for both vegetarians and nonvegetarians (Pulz and Gross, 2004; Ward and Singh, 2005; Whelan and Rust, 2006). Unlike the highly competitive marine

TABLE 10.7
Life Stage Benefits of Omega-3 EPA and DHA)

Life Stage	Key Benefits
Pregnant and nursing mothers	Good maternal health; supports fetal brain, eye, and nerve development
Infants	Brain, eye, and nerve development
Children	Part of a well-balanced diet
Teens	Part of a well-balanced diet
Adults	Cardiovascular health, mental health (Alzheimer's disease), cancer, inflammation (morning stiffness, joint pain, swelling and fatigue), and age-related disorders
Seniors	Cardiovascular health, mental health (Alzheimer's disease), cancer, inflammation (morning stiffness, joint pain, swelling and fatigue), and age-related disorders

Source: Adapted from Life stages benefits (2011), Ward and Singh (2005), Holub (2011).

fish oil market, the vegetarian omega-3 EPA and DHA source remains moderately competitive, with the dominant sources of omega-3 being algal and flaxseed oil. DHA market estimations alone are valued at US$15 million, and are considered one of the fast-growing microalgal products.

Martek Biosciences has been the dominant player in sustainable algal omega-3 technology and, as a result, this platform has given rise to products such as algae-derived omega-3 DHA in infant formula, dietary supplements, functional foods and beverages, as well as animal feed products. Martek Biosciences was recently (December 21, 2010) acquired by DSM (a global life sciences and material sciences company). The company's "flagship" product is *life's* DHA™, a completely vegetarian source of algal DHA. Martek's omega-3 technologies are secured by a robust intellectual property portfolio and supported by a strong R&D platform, with particular emphasis on the infant formula and infant nutrition area. This provides new opportunities for DSM in the infant nutrition segment, as well as in the food, beverage, and dietary supplement industries. Formulas incorporating Martek's DHA oil are available in more than sixty countries, including the United Kingdom, Mexico, China, the United States, and Canada (Spolaore et al., 2006).

10.2.3.2 DHA Production Process

Despite the strength of algal oils as products, they comprise a low share in the omega-3 ingredients market, purely because of the lack of competitors in the marketplace. Martek's microalgal cultivation occurs in fermenters, ranging from 80 to 260 m^3 in size (Ratledge, 2004; Spolaore et al., 2006; *life's* DHA website, 2012). The microalgae are grown heterotrophically utilizing glucose and yeast extracts as carbon sources (Lee, 1997). Advancing microalgal technology has been the thrust that keeps Martek highly competitive as an algal omega-3 ingredient manufacturer. *Crypthecodinium cohnii* was identified and commercially exploited by Martek as "a rich source of docosahexaenoic acid (DHA)" (Martek Corporation website, 2012), producing algal oil containing 40% to 50% DHA but no EPA (Ratledge, 2004; Ward and Singh, 2005; Spoloare et al., 2006).

Many argue that it is not imperative to consume both EPA and DHA as the human body efficiently converts EPA to DHA (Halliday, 2006). As a result, the potential new players on the algal market are focusing their efforts on increased production of high-purity EPA. The annual global demand for EPA is around 300 tonnes (Molina-Grima, 2003; Milledge, 2011). The current market value of fish oil EPA ethyl ester (95% pure) in bulk quantities is about \$650 kg^{-1} (Belarbi et al., 2000); thus, a new source such as microalgal EPA is expected to be market competitive.

10.2.3.3 EPA Production Process

At present, new players Aurora Algae and AlgaeBio (both producing biomass in autotrophic open ponds) have not yet progressed to commercialization with final products on the shelf. Aurora Algae's crude algal oil prototype contains 65% EPA and is intended for use in the pharmaceutical, animal feed, as well as heath food and beverage sectors (Aurora Algae Online, 2011). These prototypes have been distributed to potential customers and, according to Van Der Meulen, Aurora Algae has already "signed multiple letters of intent with key players across the industry" (Watson, 2011b). Bob Thompson, chairman of AlgaeBio, believes that they have a competitive economic advantage in terms of production costs. Between their patent, proprietary information, and intellectual property, they can "produce a wide array of high-value, algae-based products at a fraction of the cost" compared with their competitors (Watson, 2011b). There is still a lack of information available on the potential products in the pipeline.

The University of Almeria (Spain) has developed an outdoor tubular photobioreactor process for producing "high-purity" 96% EPA from *Phaeodactylum tricornutum*. The total cost of production of the esterified oil occurs at US\$4,602 kg^{-1}, with an estimated yield of 430 kg yr^{-1} (Molina-Grima, 2003). Some 60% of this cost is attributable to the recovery process, and the remaining 40% accounts for biomass production costs. The total cost still needs to be reduced by 80% to be economically feasible. The most common lipid extraction methods include oil press, solvent extraction, super-critical fluid extraction, and ultrasound (Harun et al., 2010). Solvent extraction is the most common method employed in the recovery of fatty acids from microalgae (Belarbi et al., 2000).

Solazyme-Roquette has created "high-lipid algal flour" (Daniells, 2011), intended for use as a main ingredient alternative to make healthier processed foods such as chocolate milk (4.5% algal flour), frozen desserts, and even low-calorie salad dressings. Household names such as Unilever, Nestle, and Abbott Laboratories are a few companies jumping onto the "omega-3 bandwagon." A fast-moving consumer goods company, Unilever has invested in a multimillion-dollar deal with Solazyme Inc. to potentially replace palm oil with algal oil as a sustainable alternative in products such as food, soaps, and lotions (Sonne, 2010).

PUFAs in general aid in the prevention and treatment of scaly dermatitis and skin dehydration (Kim et al., 2008). Ethanolic or supercritical CO_2 extracts are gaining commercial recognition in lipid-based creams and lotions as a result of their nourishing and protective effects on the skin. In progressing skin care research, glycol- and phospholipids should be given special attention (Pulz and Gross, 2004). Novel and innovative cost-effective technologies are the way to satisfy the growing demands of the health-conscious consumer.

10.2.4   OTHER POTENTIAL APPLICATIONS OF ALGAL BIOMASS

Apart from the key algal compound groupings, mentioned above, there are new market sectors and applications emerging in algal biotechnology.

10.2.4.1   Cosmetic Extracts

Marine microalgae contribute to a range of extracts rich in proteins, vitamins, and minerals, which are incorporated as active ingredients into a number of cosmetic products (Kim et al., 2008). In addition to carotenoids, phycobiliproteins, and PUFAs, microalgae produce a number of other compounds (that exhibit a range of benefits) appealing to cosmetic formulators (Table 10.8).

These compounds prevent blemishes, repair damaged skin, aid in the treatment of seborrhoea (greasy skin caused by excess sebum), and inhibit the inflammation process (Kim et al., 2008). They are formulated into face and skin care products,

TABLE 10.8
Microalgal Compounds and Their Cosmeceutical Properties

Compound	Cosmeceutical Properties
Microsporines and microsporine-like amino acids (MAAs)	Skin protection against UV radiation Antioxidant
Tocopherols	Protection against UV irradiation or oxidative damage Prevention of light-induced pathologies of the human skin and eyes Prevention of degenerative disorders (atherosclerosis, cardiovascular disease, and cancer)
Phenolic compounds	Antioxidative action Protection against UV irradiation
Terpenoids	Antioxidant Emmolient Blood stimulant Diuretic Moisturizing activities

Source: Adapted from Kim et al. (2008).

such as anti-aging creams and moisturizers, sun protection products, hair care products, refresherant or regenerant care products, emollients, and anti-irritant skin peels (Spolaore et al., 2006; Carlsson et al., 2007). *Arthrospira* (*Spirulina*) and *Chlorella* sp. are the two main genera that have established positions in the skin care market (Table 10.9).

The LVMH Group (Louis Vuitton and Moët Hennessey) (Paris, France) and Danial Jouvance (Carnac, France) have both invested in microalgal production systems (Spolaore et al., 2006; and Kim et al., 2008). It is evident that the largest market for micro- and macroalgal cosmetics is in France, with a demand estimated at 5,000 tonnes (Kim et al., 2008). This demand will continue to escalate, with the cosmetic industry evincing more interest as research and extensive studies progressively highlight the benefits of microalgal extracts on skin health.

TABLE 10.9
Cosmetic Companies Producing Commercial Products Formulated with Microalgal Extracts

Company	Commercially Exploited Microalgae	Product	Effects and Benefits
Exsymol SAM, Monaco www.exsymol.com	*Arthrospira* (*Spirulina*)	Protulines® (protein-rich extract)	Anti-aging, skin-tightening effect Prevents stria formation
Codif Recherché et Nature, St Malo, France http://www.codif-recherche-et-nature.com	*Chlorella vulgaris*	Dermochlorella® (extract)	Stimulates collagen synthesis Supports tissue regeneration Wrinkle reduction
Pentapharm, Basel Switzerland www.pentapharm.com	*Nannochloropsis oculata*	Pepha®-Tight (ingredient)	Long- and short-term skin-tightening effects
	Dunaliella salina	Pepha®-Ctive (ingredient)	Stimulate cell proliferation and turnover Positively influence skin cell metabolism
AGI Dermatics Inc. www.remergentskin.com	*Anacystis nidulans*	Remergent™ DNA Repair Formula	Repairs UV damaged skin Resists future photo-damage through natural recuperation
Company details withheld	*D. salina*	Blue Retinol™	Stimulates cell growth and proliferation
		Marestil® (Extract)	Strong moisturizing, elasticizing, and toning complex

Source: Adapted from Spolaore et al. (2006) and Kim et al. (2008).

10.2.4.2 Stable Isotope Biochemicals

Microalgae are also well suited to produce isotopically labeled compounds due to their ability to incorporate stable isotopes from inexpensive inorganic molecules into high-value isotopic organic chemicals. The ability to cultivate phototropic algae under strictly controlled conditions enables the easy incorporation of stable isotopes from inorganic carbon, hydrogen, and nitrogen sources (Pulz and Gross, 2004; Spolaore et al., 2006; Milledge, 2011). These stable isotopic compounds are used to facilitate the structural determination (at atomic level) of proteins, carbohydrates, and nucleic acids. In addition to metabolic studies (Spolaore et al., 2006), they can also be employed for clinical purposes such as gastrointestinal or breath diagnosis tests (Radmer, 1996; Pulz and Gross, 2004). Table 10.10 indicates some of the isotopically labeled microalgal products.

The market value of these compounds is estimated at US$13 million per year. A major distributor of such isotopic compounds is Spectra Stable Isotopes (Andover, MA; acquired by Cambridge Isotope Laboratories [CIL] in 2008) (Spolaore et al., 2006).

10.2.4.3 Human Nutrition

The consumption of microalgae is restricted to very few species, for example *Spirulina*, *Chlorella*, and *Dunaliella* (Jensen, 1993; Pulz and Gross, 2004). The market value of microalgal products (health foods) is estimated at US$20–25 million and it is by far the largest commercial application of microalgae (Metting, 1996). *Spirulina* and *Chlorella* are currently dominating the microalgal market. *Spirulina* is a source of protein that is comparable to meat and dairy products. *Spirulina* also contains high amounts of vitamin A and B12 (Metting, 1996).

Microalgal products to be used for human nutrition are usually sold in the form of tablets or powders (Metting, 1996; Radmer, 1996; Pulz and Gross, 2004). The packaged food industry, valued at US$2 trillion, is on the hunt for sustainable and natural sources of fiber and healthy fats as ingredients for nutritionally high-value and -quality foods (Singh et al., 1996). Microalgal foods pioneer Solazyme-Roquette has created "high-lipid algal flour" (Daniells, 2011), intended for use as a main ingredient alternative to make healthier processed foods such as chocolate milk (4.5% algal flour), frozen desserts, and even low-calorie salad dressings.

TABLE 10.10
Prices of Different Isotopically Labeled Products

Product Name	Price (US$ g^{-1})	Ref.
$^{13}C_6$-D-glucose	140	Fernández et al., 2005
^{13}C-Mixed free fatty acids	200	Spolaore et al., 2006
^{13}C-Spirulina	250	Fernández et al., 2005
^{15}N-Alanine	260	Spolaore et al., 2006
$^2H_7$, ^{13}C, $^{15}N_4$-Arginine	5,900	Spolaore et al., 2006
dATP-CN	26,000	Spolaore et al., 2006

Extracts from microalgae are creating a new sector for microalgal products (Pulz and Gross, 2004). Products made from algal extracts include *Chlorella* health drinks and *Spirulina* liquid CO_2 extracted antioxidant capsules. The microalgal biomass from *Spirulina* and *Chlorella* is not only used in human nutrition, but also in animal feed (Pulz and Gross, 2004), as it has been proven to support the immune system of animals. The market value of *Spirulina* and *Chlorella* is estimated at US\$80 and \$100 million, respectively (Radmer, 1996).

10.2.4.4 Biofertilizers

Algal biomass is the main product in microalgal technology and has various applications. The final biomass product is usually green or orange in color (Pulz and Gross, 2004). Most commercial fertilizers are derived from petroleum; however, rising fuel prices influence the cost price of commercial fertilizers derived from petroleum. A cost-effective alternative would be the use of algal biomass as organic fertilizers (http://www.algaewheel.com).

Microalgae have been used in the agriculture industry as biofertilizers and as soil conditioners (Metting, 1996). Employing microalgae as biofertilizers and soil conditioners is a common agricultural practice in Asian countries such as China and India, where they provide more than 20 kg nitrogen $ha^{-1}y^{-1}$. Nitrogen-fixing cyanobacteria such as *Anabaena, Nostoc, Aulosira Tolypothrix,* and *Scytonema* are used in rice cultivation. Mucilage-producing species of the genus *Chlamydomonas* have been used as soil conditioners to control soil erosion of pivot-irrigated soils in North America (Metting, 1996). The rationale behind using microalgae as biofertilizers is that they have the ability to increase the water-binding capacity and mineral composition of the soil (Pulz and Gross, 2004). This market generates a turnover of US\$5 billion y^{-1} (Pulz and Gross, 2004).

10.2.4.5 Bioremediation/Phycoremediation

The use of microalgae for municipal wastewater treatment has been a focus of research and development for decades as they have the ability to metabolize sewage more rapidly than bacterial treatments (Olguín, 2003). Through photosynthesis, algae assimilate nitrates, phosphates, and other nutrients present in the wastewater (http://www.algaewheel.com). In addition, the oxygen given off by algae is the primary contribution toward the treatment of municipal wastewaters and industrial effluents (Metting, 1996). Wastewater treatment systems that rely on microalgae for oxygen production are dominated by chlorophytes (Metting, 1996).

Additionally, biomass from high-rate algal pond (HRAP) systems (such as animal wastewater and fish farm wastewater) can be harvested for use as animal feed; a concept that has been demonstrated by Lincoln and Earle (1990) and Metting (1996), as part of an integrated recycling system (IRS) (Olguin, 2003). Such a system would incorporate animal waste as an input and several by-products and high-value-added products (algae) as overall outputs. "Bioespirulinema," a system carried out by Olguín (2003), has been operating effectively, and with a 4-year average *Spirulina* productivity of 39.8 tonnes $ha^{-1}y^{-1}$. The average protein content of the ash-free *Spirulina* biomass was 48.39% dry weight; which is relatively high for a system where there are no nitrogen costs.

Low harvesting costs could be one of the key concepts in establishing the economic viability of the entire system. However, these applications remain in their infancy, and extensive research and development are needed. Successful technologies and processes are available for wastewater treatment, such as the Advanced Integrated Wastewater Pond Systems (AIWPS) Technology, commercialized by Oswald and Green in the United States (Olguín, 2003). Phycoremediation with the employment of microalgae is a field with great promise and demand with so many regions in the world prone to eutrophication.

10.3   CONCLUSION

Since the use of microalgae to survive the famine in China some 2,000 years ago, the commercial applications of microalgae have been increasing rapidly. Of the many microalgal species that exist, a few species are stored in collections, and only a handful have been exploited for high-value products (Olaizola, 2003); hence, there are only a few high-value products in the marketplace (Milledge, 2011). The challenge in progressing to commercialization can be overcome by focusing efforts on products with a huge market potential and a distinct competitive advantage in large markets such as food.

Algal biomass "health food" appears to be the main commercial product, followed by food additives in the form of carotenes, pigments, and fatty acids. Algal production within the health-food market has the highest sales value but is largely dependent on health benefits and proof of efficacy (Becker, 2007; Milledge, 2011). As natural additives, these commodities are superior to synthetic products, although there is much to consider regarding the economics, sustainability, and environmental perspectives of the production of each product (Harun et al, 2010; Milledge, 2011).

There are various factors to consider in developing manufacturing processes of high-value metabolites. These include ensuring that proper taxonomic treatment is applied such that efficient screening of the microalgae can be conducted—not only for the fastest growing species, but also for those organisms with desirable robust characteristics and valuable products. A key starting point is to expand the inventory of microalgal species represented in culture collections and cell banks (Pulz and Gross, 2004; Sekar and Chandramohan, 2008). Production systems are also an important factor to consider. The type of production system depends on the nature and value of the end-product (Metting, 1996). Currently, outdoor open-pond systems are the mainstream mode of microalgal cultivation (Spolaore et al., 2006). The most successful genera cultivated in open-pond systems are *Spirulina, Dunaliella,* and *Chlorella.* Microalgal products of high value and purity, such as isotopically labeled research compounds and reagent-grade phycobilins, are produced in photobioreactor systems (Metting, 1993; Millledge, 2011). Overall operating and maintenance costs of open-pond systems are lower compared to those of photobioreactors (which are restricted mainly to the production of high-value products). Ideally, open ponds make for a competitive cultivation alternative (Harun et al., 2010) and are likely to be the way for commercial cultivation of microalgae. The location of the pond, algal strain, light and CO_2 availability, final product yield, and quality are important factors to consider in open-pond cultivation systems.

Harvesting and metabolite recovery methods depend on the nature of the species and end-product. Centrifugation is probably the most reliable method of harvesting but, on the other hand, it is costly. Filtration and flocculation are cost-effective methods that are widely used for the harvesting of algal biomass. The cost of the downstream recovery process for such high-value, high-purity products contributes to a significant portion of the overall production cost. For example, 60% of the total production cost of EPA is attributed to the recovery process of EPA (Grima et al., 2003), while biomass production only contributes approximately 40% of the total production cost. Thus, reducing the cost of downstream processing can significantly influence the overall economics of microalgal metabolite production (Grima et al., 2003).

Genetic modification of microalgae has been considered for improving the yield of valuable products at reduced costs (Milledge, 2011). The production of recombinant proteins in microalgal chloroplasts has several attributes (Specht, 2010). Transgenic proteins can accumulate to much higher levels in the chloroplasts than when expressed from the nuclear genome; chloroplasts can be transformed with multiple genes in a single event due to multiple insertion sites (Specht, 2010). Furthermore, proteins produced in chloroplasts are not glycosylated (Franklin and Mayfield, 2005); this can be useful in the production of antibodies that are similar to native antibodies in their ability to recognize their antigens (Specht, 2010). To demonstrate the feasibility of human antibody expression in an algal system, a full-length IgG (Immunoglobulin G) antibody has been synthesized in the chloroplast of the green alga *Chlamydomonas reinhardtii* (Hempel et al., 2011). The ability to accumulate high-value compounds makes microalgae attractive for recombinant protein production; however, there are some factors that limit microalgal expression systems (Gong et al., 2011). These include the lack of standard procedures for genetic transformation of commercially important microalgal species, limited availability of molecular toolkits for genetic modification of microalgae, and low expression levels of recombinant proteins (Surzycki et al., 2009).

The use of genetic modification may reduce the organic and natural appeal of specific algal products, especially when the product is to be applied in the food and feed industries. It is thus imperative to prioritize endeavors toward proper species selection and production process development. This is a preferred approach, rather than resorting to genetic engineering of microalgae. However, for specialized applications, such as for therapeutic and diagnostic purposes, the use of microalgae as bioreactors for the production of recombinant proteins may be advantageous.

Microalgae boast a range of high-purity, valuable products that have progressed successfully to commercialization in applications in the food, pharmaceutical, clinical research, and animal nutrition industries. The possible employment of microalgae in environmental applications (phycoremediaion and biofertilizers) provides potential solutions to global warming and sustainable economic development. Although in their infancy, these applications hold significant promise, and with potential use in diagnostics and therapeutics, the range of applications continues to grow. However, for these industries to progress, it is important to start at grass-root levels in research. Exhaustive screening procedures must be conducted for specific species and products, while also considering the economics of upstream and downstream processes for individual products.

REFERENCES

Algaewheel.com. Online. Available at <http://www.algaewheel.com> (Accessed 4 April 2012).

Arteburn, L.M., Boswell, K.D., Henwood, S.M., and Kyle, D.J. (2000). A developmental study in rats using DHA- and ARA-rich single-cell oils. *Food and Chemical Toxicology,* 38: 763–771.

Aurora Algae (2011). Make Way for a Better Omega Source. Online. Available at <http://www.aurorainc.com/solutions/omega-3/> (Accessed 7 December 2011).

Becker, E.W. (2007). Microalgae as a source of protein. *Biotechnology Advances,* 25: 207–210.

Berlabi, E.H., Molina, E., and Chisti, Y. (2000). A process for high yield and scaleable recovery of high purity eicosapentaenoic acid esters from microalgae and fish oil. *Enzyme and Microbial Technology,* 26: 516–529.

Borowitzka, M.A. (1992). Algal Biotechnology products and processes—Matching science and economics. *Journal of Applied Phycology,* 4: 267–279.

Carlsson, A.S., Van Beilen, J.B., Möller, R., and Clayton, D. (2007). Micro- and macro-algae: Utility for industrial applications. In *EPOBIO: Realising Economic Potential of Sustainable Resources—Bioproducts from Non-Food Crops.* Diana Bowles (Ed.), CPL Press, Tall Gables, The Sydings, Speen, Newbury, Berks RG14 1RZ, UK.

Cosgrove, J. (2010). The Carotenoid Market: Beyond beta-Carotene. Online. Available at <http://www.nutraceuticalsworld.com/contents/view_online-exclusives/2010-12-13/the-carotenoid-market-beyond-beta-carotene/> (Accessed 26 January 2012).

Daniells, S. (2011). Solazyme-Roquette's Algal Flour Promises Exciting Future for Delicious, Low-Fat Food. Online. Available at <http://www.foodnavigator-usa.com/Business/Solazyme-Roquette-s-algal-flour-promises-exciting-future-for-delicious-low-fat-food> (Accessed 17 February 2012).

Del Campo, J.A, Garcia-González, M., and Guerrero, M.G. (2007). Outdoor cultivation of microalgae for carotenoid production: current state and perspectives. *Applied Microbiology and Biotechnology,* 74: 1163–1174.

Dufossé, L., Galaup, P., Yaron, A., Arad, S.M., Blanc, P., Chidambara Murthy, K.N., and Ravishankar, G.A. (2005). Microorganisms and microalgae as pigments for food use: A scientific oddity or an industrial reality? *Trends in Food Science and Technology,* 16: 389–406.

Eriksen, N.T. (2008). Production of phycocyanin-*a* pigment with applications in biology, biotechnology, foods and medicine. *Applied Microbiology and Biotechnology,* 80: 1–14.

Fernández, F.G.A., Fernández Sevilla, J.M., Egorova-Zachernyuk, T.A., and Molina Grima, E. (2005). Cost effective production of ^{13}C and ^{15}N stable isotope-labelled biomass from phototrophic microalgae for various biotechnological applications. *Biomolecular Engineering,* 22: 193–200.

Franklin, S.E., and Mayfield, S.P. (2005). Recent developments in the production of human therapeutic proteins in eukaryotic algae. *Expert Opinion on Biological Therapy,* 5: 225–235.

Freeman, M.P., Hibbeln, J.R., and Wisner, K.L. (2006). Omega-3 fatty acids: Evidence basis for treatment and future research in psychiatry. *Journal of Clinical Psychiatry,* 67: 1954–1967.

Gong, Y., Hu, H., Gao, Y., Xu, X., and Gao, H. (2011). Microalgae as platforms for production of recombinant proteins and valuable compounds: Progress and prospects. *Journal of Industrial Microbiology and Biotechnology,* 38: 1879–1890.

Grima, E.M., Berlabi, E.-H., Acién Fernández, F.G., Robles Medina, A., and Chisti, Y. (2003). Recovery of microalgal biomass and metabolites: process options and economics. *Biotechnology Advances,* 20: 491–515.

Guedes, A.C., Amaro, H.M., and Malcata, F.X. (2011). Microalgae as sources of carotenoids. *Marine Drugs,* 9: 625–644.

Halliday, J. (2006). Swiss Company Derives DHA and EPA from Algae. Online. Available at <http://www.nutraingredients.com/Industry/Swiss-company-derives-DHA-and-EPA-from-algae> (Accessed 29 July 2011).

Harun, R., Singh, M., Ford, G.M., and Danquah, M.K. (2010). Bioprocess engineering of microalgae to produce a variety of consumer products, *Renewable and Sustainable Energy Reviews,* 14: 1037–1047.

Heller, L. (2008). ZMC takes up position in lutein, zeaxanthin market. Online. Available at <http://www.nutraingredients-usa.com/Industry/ZMC-takes-up-position-in-lutein-zeaxanthin-market> (Accessed 05 April 2012).

Hempel, F., Lau, J., Klingl, A., and Maier, U.G. (2011). Algae as protein factories: Expression of a human antibody and the respective antigen in the diatom *Phaeodactylum tricornutum,* *Plos One,* 6: 1–7.

Holub, B.J. (2011). DHA/EPA Omega-3 Fatty Acids for Human Health and Chronic Disorders Insider Ingredients Volume 1: Issue 1. Online. Available at <http://www.naturalproductsinsider.com/lib/download/asset-exploring-the-evolving-market-for-long-chain-omega-3s.ashx?item_id={A9349DEE-989E-4AC5-B8FE-F30D34078CEE}&item_name=digital_issue-exploring-the-evolving-market-for-long-chain-omega-3s> (Accessed 29 July 2011).

Inbaraj, B.S., Chien, J.T., and Chen, B.H. (2006). Improved high performance liquid chromatographic method determination of carotenoids in the microalga *Chlorella pyrenoidosa.* *Journal of Chromatography A,* 1102: 193–199.

Jensen, A. (1993). Present and future needs for algae and algal products. *Hydrobiologia,* 260/261: 15–23.

Kim, S., Ravichandran, Y.D., Khan, S.B., and Kim, Y.T. (2008) Prospective of the cosmoceuticals derived from marine organisms. *Biotechnology and Bioprocess Engineering,* 13: 511–523.

Kotake-Nara, E., and Nagao, A. (2011). Absorption and metabolism of xanthophylls. *Marine Drugs,* 9: 1024–1037.

Kris-Etherton, P.M., Harris, W.S., and Appel, L.J. (for the AHA Nutrition Committee) (2003). Omega-3 fatty acids and cardiovascular disease—New recommendations from the American Heart Association. *Arteriosclerosis, Thrombosis, and Vascular Biology,* 23: 151–152.

Lee, Y.-K. (1997). Commercial productions of microalgae in the Asia-Pacific Rim. *Journal of Applied Phycology,* 9: 403–411.

Lincoln, E.P. and Earle, J.F.K. (1990) Wastewater treatment with micro-algae. In: Akatsuka, I. (Ed.), *Introduction to Applied Phycology,* SSPB Academic Publishing, The Hague, pp. 429–446.

Luiten, E.E.M., Akkerman, I., Koulman, A., Kamermans, P., Reith, H., Barbosa, M.J., Sipkema, D., and Wijffels, T.H. (2003). Realising the promises of marine biotechnology. *Biomolecular Engineering,* 20: 429–439.

life's DHA™ website (2012). Online. Available at <http://www.lifesdha.ca/-em-lifesdha-em.aspx> (Accessed 9 March 2012).

Life Stages Benefits (2011). Meg-3. Online. Available at <http://www.meg-3.com/lifestages-benefits> (Accessed 30 June 2011).

Market Opportunities (2011). Ocean Nutrition Canada. Online. Available at <http://www.ocean-nutrition.com/omega3/market_opportunities> (Accessed 8 July 2011).

Martek Corporation Site (2012). History of Martek—Visionary Plant Based Nutrition Research. Online. Available at <http://www.martek.com/About/History.aspx> (Accessed 13 February 2012).

Metting, Jr., F.B. (1996). Biodiversity and application of microalgae. *Journal of Industrial Microbiology,* 17: 477–489.

Milledge, J.J. (2011). Commercial application of microalgae other than as biofuels: A brief review. *Reviews in Environmental Science and Biotechnology,* 10: 31–41.

Mischoulon, D., Best-Popescu, C., Laposata, M., Merens, W., Murakami, J.L., Wu, S.L., Papakostas, G.I., Dording, C.M., Sonawalla, S.B., Nierenberg, A.A., Alpert, J.E., and Fava, M. (2008). A double-blind dose-finding pilot study of docosahexaenoic acid (DHA) for major depressive disorder. *European Neuropsychopharmacology,* 18: 639–645.

Molina-Grima, E., Berlabi, E.-H., Fernandez, F.G.A., Medina, A.R., and Chisti, Y. (2003). Recovery of microalgal biomass and metabolites: Process options and economics, *Biotechnology Advances,* 20: 491–515.

Muller-Feuga, A. (2000). The role of microalgae in aquaculture: Situation and trends. *Journal of Applied Phycology,* 12: 527–534.

Murthy, K.N.C., Rajesha, J., Mahadeva, S., and Ravishankar, G.A. (2005). Comparative evaluation of hepatoprotective activity of carotenoids in microalgae. *Journal of Medicinal Food,* 8: 523–528.

Nemets, B., Stahl, Z., and Belmaker, R.H. (2002). Addition of omega-3 fatty acids to maintenance medication treatment for recurrent unipolar depressive disorder. *American Journal of Psychiatry,* 159: 477–479.

Ocean Nutrition Canada (2011). Market Opportunities. Online. Available at <http://www.ocean-nutrition.com/omega3/market_opportunities> (Accessed 8 July 2011).

Olaizola, M. (2003). Commercial development of microalgal biotechnology: From the test tube to the marketplace. *Biomolecular Engineering,* 20: 459–466.

Olguín, E.J. (2003). Phycoremediation: Key issues for cost-effective nutrient removal processes. *Biotechnology Advances,* 22: 81–91.

Orosa, M., Torres, E., Fidalgo, P., and Abalde, J. (2000). Production and analysis of secondary carotenoids in green algae. *Journal of Applied Phycology,* 12: 553–556.

Pulz, O., and Gross, W. (2004). Valuable products from biotechnology of microalgae. *Applied Microbiology and Biotechnology,* 65: 635–648.

Radmer, R.J. (1996). Algal diversity and commercial algal products. *Marine Biotechnology,* 46: 263–270.

Ratledge, C. (2004). Fatty acid biosynthesis in microorganisms being used for single cell oil production. *Biochimie,* 86: 807–815.

Ruxton, C.H.S., Reed, S.C.S., and Millington, K.J. (2004). The health benefits of omega-3 fatty acids: A review of the evidence. *Journal of Human Nutrition and Diet,* 17: 449–459.

Schmidt-Dannert, C., Umeno, D., and Arnol, F.H. (2000). Molecular breeding of carotenoid biosynthetic pathways. *Nature Biotechnology,* 18: 750–753.

Sekar, S., and Chandramohan, M. (2008). Phycobiliproteins as a commodity: Trends in applied research, patents and commercialization. *Journal of Applied Phycology,* 20: 113–136.

Simopoulos, A.P. (1999). Essential fatty acids in health and chronic disease. *The American Journal of Clinical Nutrition,* 70 (Suppl.): 560S–569S.

Singh, S., Bhushan, K.N., and Banerjee, U.C. (2005). Bioactive compounds from cyanobacteria and microalgae: An overview. *Critical Review Biotechnology,* 25: 73–95.

Solazyme (2012). Market Areas/Overview. Online. Available at <http://solazyme.com/market-areas> (Accessed 21 February 2012).

Sonne, P. (2010). To wash hands off palm oil Unilever embraces algae. Dow Jones and Company website. *The Wall Street Journal.* Online. Available at <http://online.wsj.

com/article/SB10001424052748703720004575477531661393258.html> (Accessed 15 February 2012).

Specht, E., Miyake-Stoner, S., and Mayfield, S. (2010). Micro-algae come of age as a platform for recombinant protein production. *Biotechnology Letters,* 32: 1373–1383.

Spolaore, P., Joannis-Cassan, C., Duran, E., and Isambert, A. (2006). Commercial applications of microalgae. *Journal of Bioscience and Bioengineering,* 101(2): 87–96.

Surzycki, R., Greenham, K., Kitayama, K., Dibal, F., Wagner, R., Rochaix, J.-D., Ajam, T., and Surzycki, S. (2009). Factors affecting expression of vaccines in microalgae. *Biologicals,* 37: 133–138.

Takaichi, S. (2011). Carotenoids in algae: Distributions, biosyntheses and functions, *Marine Drugs,* 9: 1101–1118.

U.S. DOE (2010). National Algal Biofuels Technology Roadmap. U.S. Department of Energy, Office of Energy Efficiency and Renewable Energy, Biomass Program. Online. Available at <http://biomass.energy.gov> (Accessed 31 January 2012).

Vital solutions (2010) Algal Technology platform market research. Online. Available at <http://www.vitalsolutions. Biz/cms/home.html> (Accessed 1 February 2012).

Von Schacky, C., and Harris, W.S. (2007). Cardiovascular benefits of omega-3 fatty acids. *Cardiovascular Research,* 73: 310–315.

Ward, O.P., and Singh, A. (2005). Omega-3/6 fatty acids: Alternative sources of production. *Process Biochemistry,* 40: 3627–3652.

Watson, E. (2011a). Algal omega-3 market heats up as new players bid for a slice of the action. Online. Available at <http://www.nutraingredients-usa.com/Industry/Algal-omega-3-market-heats-up-as-new-players-bid-for-a-slice-of-the-action> (Accessed on September 2011).

Watson, E. (2011b). Aurora Algae ups ante in omega-3 market plans for Q4, 2012 launch NUTRA Ingredients-USA Special Edition. Online. Available at <http://www.nutrain-gredients-usa.com/Industry/Aurora-Algae-ups-ante-in-omega-3-market-with-plans-for-Q4-2012-launch > (Accessed on 23 September 2011).

Weiss, A., Johannisbauer, W., Gutsche, B., Martin, L., Cordero, B.F., Rodriguez, H., Vargas, A.M., and Obrastzova, I. (2008). Process for Obtaining Zeaxanthin from Algae, Patent Number EP1806411.

Wen, Z., and Chen, F. (2003). Heterotrophic production of eicosapentaenoic acid by microalgae, *Biotechnology Advances,* 21: 273–294.

Whelan, J., and Rust, C. (2006). Innovative dietary sources of n-3 fatty acids. *Annual Review of Nutrition,* 26: 75–103.

11 Algae-Mediated Carbon Dioxide Sequestration for Climate Change Mitigation and Conversion to Value-Added Products

Ajam Y. Shekh, Kannan Krishnamurthi,
Raju R. Yadav, Sivanesan S. Devi, Tapan
Chakrabarti, and Sandeep N. Mudliar
CSIR-National Environmental Engineering
Research Institute (NEERI)
Nagpur, India

Vikas S. Chauhan and Ravi Sarada
CSIR-Central Food Technological Research Institute (CFTRI)
Mysore, India

Sanniyasi Elumalai
Presidency College
Chennai, India

CONTENTS

11.1 INTRODUCTION

The increase in the atmospheric concentration of carbon dioxide (CO_2) due to anthropogenic interventions has led to several undesirable consequences, which include increasing Earth temperature, violent storms, melting of polar ice sheets, and sea level elevations (Shekh et al., 2012). In the global effort to combat and mitigate climate change, several CO_2 capture and storage technologies are being deliberated. Some of the CO_2 abatement processes currently in use include the use of chemical/physical solvents, adsorbents onto solids, membranes, cryogenic/condensation systems, and geological and deep ocean sequestration (Abu-Khader, 2006; Shekh et al., 2012; Yadav et al., 2012). In practice, the above-mentioned approaches are questionable with respect to their cost effectiveness (Abu-Khader, 2006; Shekh et al., 2012). Therefore, there is an urgency to look for sustainable, economical, and replicable technologies for CO_2 sequestration. Microalgae have attracted a great deal of attention for CO_2 fixation because of their ability to convert CO_2 into biomass via photosynthesis at much higher rates than conventional terrestrial land-based crops (Chisti, 2007; 2008). Microalgae are able to grow on agriculturally nonproductive arid lands, in saline water, and in domestic and industrial wastewaters, and consequently do not compete with conventional food crops grown on agricultural land and thus pose no threat to food security issues (Sheehan et al., 1998).

Similarly, *Dunaliella* is gaining popularity as a source of β-carotene. *Haematococcus* is being grown for the production of the ketocarotenoid Astaxanthin. Further, *Botryococcus* species are a promising renewable energy source as they accumulate very large quantities of hydrocarbons (30% to 73% of dry weight) and also have a high octane rating as a fuel source because of their highly branched structures. Therefore, one of the most promising future-proof CO_2 sequestration technologies may be microalgal cultivation integrated with CO_2 sequestration and its conversion to value-added food and fuel-grade precursors/products. This chapter deliberates on some of these aspects.

11.2 MICROALGAE FOR CO_2 SEQUESTRATION: CONCEPT AND RECENT DEVELOPMENTS

The urgent need for substantive net reductions in CO_2 emissions into the atmosphere can be addressed via biological CO_2 mitigation (Ramanan et al., 2009a, b; Fulke et al., 2010; Shekh et al., 2012; Yadav et al., 2012), coupled with a transition to value-added products (VAPs) such as biofuels (Fulke et al., 2010; Kumar et al., 2010). Microalgae can fix CO_2 from the atmosphere, from flue gases, or directly as soluble carbonates by the process of photosynthesis using solar energy (Wang et al., 2008). Concurrently, biomass is produced with 10 to 15 times greater efficiency than terrestrial plants, which has application in carbon credit programs (Lam and Lee, 2011). Microalgal cells contain approximately 45% to 65% carbon, wherein 1 kg dry biomass is produced by fixing approximately 1.8 kg CO_2 (Chisti, 2007). CO_2 from the external atmosphere (air/extracellular surroundings of microalgae) can be dissolved as bicarbonates and made available to microalgae for uptake and intracellular conversion to CO_2 by intracellular carbonic anhydrases. CO_2 is then made available to Ribulose-1,5-bisphosphate carboxylase/oxygenase (RuBisCO) for its fixation into energy compounds (Kaplan et al., 1991). Microalgae may provide a better tool for simultaneous CO_2 sequestration and biofuel generation. Current CO_2 levels (0.0387% (v/v)) in the atmosphere are inefficient in supporting the high microalgal growth rates and biomass productivities needed for full-scale biofuel production (Kumar et al., 2010). Flue gases from various industries typically contain CO_2 in the concentration range around 15% (v/v), which will provide sufficient amounts of CO_2 for large-scale microalgae biomass production (Kumar et al., 2010). Owing to the cost of upstream separation of CO_2 gas, direct utilization of power plant flue gas would be advantageous in microalgal biofuel production systems. Flue gases that contain CO_2 concentrations ranging from 5% to 15% (v/v) have been scrubbed for direct use in microalgal culture systems for biomass growth (Kumar et al., 2010). This approach is believed to be pragmatic, more eco-friendly, and technologically feasible for bio-mitigation of CO_2 as compared to physicochemical adsorbents or deep-ocean injections. This is a win-win scenario wherein combating air pollution through microalgal cultivation is possible while simultaneous microalgal biomass generation can be exploited to produce biofuel and other VAPs.

A comparative evaluation of CO_2 sequestration potential of various microalgal species is presented in Table 11.1. Some microalgal species such as *Chlorella*, *Scenedesmus*, and *Botryococcus* are among the microalgae that have been studied for CO_2 consumption and are promising for bio-mitigation of CO_2 (Griffith and Harrison, 2009; Fulke et al, 2010). *Scenedesmus obliquus* was found to tolerate high CO_2 concentrations (up to 12% v/v) with optimal removal efficiency of 67%, when grown at pilot scale using industrial flue gas as a carbon source (Li et al., 2011). Biomass generation through CO_2 sequestration and exploitation of biomass for biodiesel precursor formation has been studied by Fulke et al. (2010). *Chlorella* sp. was found to have biomass productivity of 0.322 g $L^{-1}d^{-1}$ with lipid productivity of 0.161 g d^{-1} at 3% CO_2 as feed gas.

The presence of FAMEs (fatty acid methyl esters) suitable for biodiesel (e.g., palmitic acid (C 16:0), docosapentaenoic acid (C 22:5), and docosahexaenoic acid (C 22:6)) have been confirmed. The calcite produced was characterized by Fourier transform infrared (FTIR) spectroscopy, scanning electron microscopy (SEM), and x-ray diffraction (XRD) (Fulke et al., 2010). The ability to tolerate CO_2 concentration during growth is confined to the individual specie's characteristics. However, when exposed, the CO_2 concentration in the gaseous phase does not provide a true reflection of the actual concentration of CO_2 in the flue gas to which the microalgal specie is exposed during dynamic liquid suspension. It depends on the alkalinity (pH) and the CO_2 concentration gradient created by the resistance to mass transfer (Kumar et al., 2010).

11.3 MICROALGAE: VALUE-ADDED PRODUCTS (VAPS)— FUEL-BASED

Microalgae appear to be the only source of biodiesel that have the potential to completely replace fossil diesel (Table 11.2). Unlike other oil crops, microalgae grow rapidly and many of them are exceedingly rich in oil (Griffith and Harrison, 2009). Microalgae commonly double their biomass within 18 to 24 h (Sheehan et al., 1998).

TABLE 11.1

CO_2 Biofixation and Biomass Productivity of Various Microalgae in Different Reactor Configurations

Microalgal Species	CO_2 Feed Gas (%)	CO_2 Fixation Rate (g m^{-3}h^{-1}) or Removal Efficiency (%)	Specific Growth Rate[d] (h^{-1}) or Biomass Productivity[e] (g m^{-3}h^{-1})	Reactor Type
Chlamydomonas reinhardtii	30*	NA	0.08–0.01[d]	Batch hotobioreactors
Chlorella pyrenoidosa	100*	NA	0.09–0.09[d]	
Chlorogleopsis sp.	5	0.8–1.9[a]	0.0007–0.0060[d]	
Scenedesmus obliquus	60*	NA	0.06–0.04[d]	
Spirulina platensis	0.03	NA	0.0082–0.002[d]	
C. kessleri	18	NA	0.84[d]	Open photobioreactor
Chlorella sp.	6–8	10–50%	NA	
Chlorella sp.	10	46%	0.09[f]	
N. salina	5	NA	1.25[c]	
N. salina	15	NA	4.1[c]	
Spirulina LEB18	6	0.21[g]	0.2475[f]	Closed photobioreactor
Spirulina platensis	10	39%	0.1164[f]	
Botryococcus braunii	2–3	3–18[a]	NA	
Chlorella sp.	0.03	1.3784[h]	0.7511[f]	
Chlorella sp.	0.03	NA	0.099[f]	
	3	NA	0.212[f]	
	10	NA	0.045[f]	
	15	NA	0.030[f]	

Organism				Membrane photobioreactor
Chlorella vulgaris	1	128[b] and 141[b]	NA	
Euglena gracilis	11	3.1[a]	4.8[e]	
Porphyridium sp.	2–3	3–18[a]	NA	
S. obliquus AS-6-1	20	290.2[i]	150[j]	
S. obliquus CNW-N	20	390.2[i]	201.4[j]	
C. vulgaris	1	NA	4[e]	
C. vulgaris	1 & 0.04	80–260	NA	
C. vulgaris	1	43[b] and 275[b]	NA	
C. vulgaris	0.045	148[a]	NA	
Nannochloropsis	1	NA	4.2–5.8[e]	
		NA	0.8–41.7[c]	
S. platensis	2–15	38.3–60[c]	3–17.8[e]	

Source: Table is updated version of published work of Kumar, A. et al. (2010); Fulke, A. et al. (2010); Ramanan, R. et al. (2009); Zhao, B. et al. (2011); Ho, S.H. et al. (2010); Cheng, L. et al. (2006).

Note: Abbreviation: NA, not available.

a,b,c From Kumar A. et al., 2010.

d Specific growth rate (h⁻¹).

e Biomass productivity (g m⁻³/d⁻¹).

f Biomass productivity, calculated (g L⁻¹d⁻¹).

g Calculated CO_2 fixation rate (g g⁻¹d⁻¹).

h CO_2 fixation rate (g L⁻¹d⁻¹).

i CO_2 fixation rate (mg L⁻¹d⁻¹).

j Biomass productivity (mg L⁻¹d⁻¹).

*mM in medium.

TABLE 11.2

Comparison of Some Biodiesel Sources

Crop	Oil Yield (L ha^{-1}y^{-1})
Corn	172
Soybean	446
Canola	1,190
Jatropha	1,892
Oil palm	5,950
Microalgae[a]	136,900
Microalgae[b]	58,700

[a] 70% oil (by wt.) in biomass.
[b] 30% oil (by wt.) in biomass.
Source: Adapted from Chisti (2007) and Mata et al. (2010).

The oil content in microalgae can exceed 80% by weight of dry biomass (Spolaore et al., 2006). The biofuel production potentials of various algal strains reported are summarized in Table 11.3. Depending on the species, microalgae produce many different kinds of lipids, hydrocarbons, and other complex oils. Hexadecanoic acid methyl ester (16:0), palmitoleic acid methyl ester (16:1), octadecanoic acid methyl ester (18:1), and stearic acid methyl ester (18:0) are some of the major FAMEs found to be suitable for biodiesel production derived from microalgal lipids (Dayananda et al., 2007; Francisco et al., 2010; Fulke et al., 2010). Using microalgae to produce biodiesel will not compromise the production of food, fodder, and other products derived from crops (Griffith and Harrison, 2009). The strain *Botryococcus braunii*, however, grows slowly and produces about 30% to 73% hydrocarbons under laboratory conditions (Dayananda et al., 2006; 2007; 2010).

For cost-effective commercial biodiesel production, appropriate strain selection according to the suitability for site of cultivation and local environmental conditions is imperative (Sheehan et al., 1998; Griffith and Harrison, 2009; Chanakya et al., 2012). The key challenge for microalgal biodiesel production is the screening and selection of microalgal species that can maintain a high growth rate with high lipid content in addition to a high metabolic rate (Griffith and Harrison, 2009). The species that are metabolically rigorous can tolerate high concentrations of salt, CO_2, high alkalinity, and high temperature; and have the ability to grow and replicate under nutritional stress by altering their metabolic pathways—these are the species that are found to be most promising in this regard (Verma et al., 2010). Nitrogen limitations have been found to enhance lipid accumulation in the microalgae (Griffith and Harrison, 2008). Yeesang and Cheirsilp (2011) studied the effect of nitrogen deprivation and iron (Fe^{3+}) enhancement with higher light intensity on lipid content. They observed an increase in lipid content from 25.8% to 35.9% (Yeesang and Cheirsilp, 2011). The findings of Liu et al. (2008) also confirmed that lipid content in *Chlorella vulgaris* increased by three- to sevenfold when the growth medium was supplemented with 0.012 mM Fe^{3+}.

TABLE 11.3
Oil Content of Some Selected Microalgae

Sr. No.	Microalgae	Oil Content (% Dry Wt)	Volumetric Productivity of Biomass (g L⁻¹d⁻¹)
1	*Botryococcus braunii*	25–75	0.02
2	*Chlorella emersonii*	29	0.036–0.041
3	*Chlamydomonas reinhardtii*	21	–
4	*Chlorella minutissima*	31	–
5	*Chlorella protothecoides*	13	2–7.70
6	*Chlorella pyrenoidosa*	18	2.90–3.64
7	*Chlorella sorokiniana*	16	0.23–1.47
8	*Chlorella vulgaris*	25	0.02–2.5
9	*Crypthecodinium cohnii*	20	10
10	*Cylindrotheca* sp.	16–37	–
11	*Dunaliella primolecta*	23	0.09
12	*Dunaliella salina*	19	0.22–0.34
13	*Dunaliella tertiolecta*	15	0.12
14	*Euglena gracilis*	20	7.70
15	*Isochrysis* sp.	25–33	0.08–0.17
16	*Monallanthus salina* N	20	0.08
17	*Nannochloris* sp.	20–35	0.17–0.51
18	*Nannochloropsis* sp.	31–68	0.17–1.43
19	*Neochloris oleoabundans*	35–54	–
20	*Nitzschia* sp.	45–47	–
21	*Phaeodactylum tricornutum*	20–30	0.003–1.9
22	*Schizochytrium* sp.	50–77	–
23	*Tetraselmis sueica*	15–23	0.12–0.32

Source: Adapted from Griffiths and Harrison (2009); Mata, et al. (2010); and Chisti (2007).

Enhancement of CO_2 sequestration and lipid accumulation is one of the major challenges that can be duly addressed by an extensive search for the new genes involved in the process (bio-prospecting) or targeted genetic engineering, both of which are promising approaches (Kumar et al., 2010).

Genetic and metabolic engineering transformations in microalgae are limited to very few microalgal species. The use of molecular biology techniques as a toolkit to engineer microalgae for biodiesel production is a demanding strategy. Understanding, incorporation, and expression of the gene encoding rate-limiting enzyme of inorganic carbon uptake and lipid biosynthetic pathways are of more importance (Badger and Price, 2003; Verma et al., 2010). With the advancements in genome sequencing with sequence availability of *Anabaena, Ostreococcus tauri, Thalassiosira pseudonana,* and other algal species (Beer et al., 2009; Verma et al., 2010), the genetic transformation of microalgal species for various purposes is now promising. *Cyclotella cryptica* and *Navicula saprophila* were genetically transformed with the acetyl-CoA carboxylase (acc) gene isolated from *Cyclotella cryptica* for enhanced lipid synthesis. Such efforts could successfully enhance the activity of the acc gene; however, no significant lipid content was found to increase in transgenic species, indicating that acc activity by itself cannot increase lipid biosynthesis and accumulation (Dunahay et al., 1996). More holistic approaches were forwarded for lipid enhancement through genetic engineering. Studies on the insights of various regulatory steps of the lipid biosynthetic pathway (Courchesne et al., 2009), expression, and regulatory analysis of genes and enzymes (such as fatty acid synthase, acetyl-CoA carboxylase, acyl-CoA, diacylglycerol acyltransferase) involved in triacylglycerol (TG) formation have been carried out (Bouvier-Nave et al., 2000; Dehesh et al., 2001; Jako et al., 2001).

Genetic transformations, which influence TG biosynthesis, may enhance biodiesel production in transgenic microalgae (Verma et al., 2010). There have been considerable enhancements in the genetic engineering aspects of algae to improve the performance of transgenic microalgae, including (1) the efficient expression of transgenes, (2) riboswitches for gene regulation in algae, (3) inducible nuclear promoters and reporter genes (luciferase), as well as (4) inducible chloroplast gene expression (Beer et al., 2010).

Transcription-level regulations by transcription factors can also be used as a strategy to control the overall metabolite flux. The effect of transcription regulatory proteins has also been studied with respect to their expression levels to increase the production of secondary metabolites of interest in plants (Verma et al., 2010). In addition to the approaches discussed above, further genome sequencing efforts need time. Advancements in existing tools and the development of new genetic transformation tools and screening methods will add further rigor to the efforts to optimize the accumulation of lipid and/or other metabolites alongside improving the economics of its production (Beer et al., 2010; Verma et al., 2010). Looking at the current interest in microalgae-based biofuels and microalgae/prototrophs, fundamental research will indisputably provide further advances in the near future (Beer et al., 2010). However, with respect to the utilization of genetically modified crops in India, that country has already accepted the release of *Bacillus thuringiensis* (Bt) cotton, which is successfully growing without causing any environmental problems. We (the authors) are of the opinion that in the near future, the scientific community will be exploring genetically modified microalgae in both open ponds as well as closed photobioreactors. But prior to doing that, several scientific issues should be addressed, and risk assessment (to ecosystem) studies must be performed to determine the legitimacy of using genetically modified microalgal strains to produce biodiesel.

11.4 MICROALGAE AS A SOURCE OF VALUE-ADDED FOOD SUPPLEMENTS

11.4.1 β-CAROTENE

The biomass of certain microalgae could find application as food supplements due to their nutritional content other than proteins. The β-carotene content of *Dunaliella salina,* a halotolerant green microalga, can reach up to 14% of dry weight (Metting, 1996). β-Carotene, a component of the photosynthetic reaction center, accumulates as lipid globules in interthylakoid spaces of chloroplasts of alga (Vorst et al., 1994). It contributes to light harvesting and protects the alga from oxidative damage during excessive irradiance by quenching the triplet-state chlorophyll or by reacting with singlet oxygen ($^1O_2$), thus preventing the formation of reactive oxygen species (Demming-Adams and Adams, 2002; Del Campo et al., 2007; Raja et al., 2007; Telfer, 2002). The beneficial effects of β-carotene on human health are attributed to its antioxidant properties (Guerin et al., 2003; Higuera-Ciapara et al., 2006; Hussein et al., 2006), and several studies have indicated that adequate intake of carotenoids has the ability to prevent degenerative diseases (Astorg, 1997; Demming-Adams and Adams, 2002; Krinsky and Johnson, 2005). β-Carotene also has the ability to act as provitamin A (Garcia-Gonzalez et. al., 2005; Gouveia and Empis, 2003). Because of these properties, β-carotene has found applications as a food supplement and colorant.

The extent of β-carotene accumulation in *Dunaliella* biomass is a function of high salinity, temperature stress, high light intensity, and nitrogen limitation. Being an extremophile, by virtue of its ability to grow at high salinity, it is possible to grow *Dunaliella* biomass in open-pond cultivation systems in photo-autotrophic mode. The production ponds are typically located in areas that could provide high solar irradiance, warm temperatures, and hypersaline waters (Ben-Amotz, 1999). Commercial cultivation facilities of *Dunaliella* are located in Australia, Israel, China, and the United States (Del Campo et al., 2007), with global production estimates at about 1,200 MT y^{-1} (Pulz and Gross, 2004). The open-pond cultivation systems used are either very large ponds (without mixing) of up to 250 ha or paddle-mixed raceway ponds of about 3,000 m^2 surface area (Del Campo, 2007). Commercial producers are offering *Dunaliella* biomass directly as a powder for application as an ingredient in human dietary supplements and functional foods (Spolaore et al., 2006).

Downstream processing of *Dunaliella* biomass is carried out to extract β-carotene for use as a natural food colorant and food supplement. The natural β-carotene from *Dunaliella* must compete with cheaper synthetic β-carotene in the marketplace. Synthetic β-carotene is dominated by all-*trans*-β-carotene (Von Laar et al., 1996), whereas natural β-carotene from *Dunaliella* contains more than 50% 9-*cis*-β-carotene (Johnson et al., 1996). Therefore, although more expensive, natural β-carotene provides the natural isomers in their natural ratio (Guerin et al., 2003; Garcia-Gonzalez et al., 2005; Spolaore et al., 2006), and the natural isomer of β-carotene is accepted as superior to the synthetic all-*trans*-isomer (Radmer, 1996; Vilchez et al., 1997; Lorenz and Cysewski, 2000; Becker, 2004; Spolaore et al., 2006). Although not yet cost compared to synthetic β-carotene, production of natural β-carotene from *Dunaliella* has been reported as an economically viable and growing industry (Singh et al., 2005; Chisti, 2006). The algal meal of *Dunaliella* after extraction of β-carotene is reported to contain about 40% protein and therefore could find application in fish and poultry feed (Iwamoto, 2004).

11.4.2 ASTAXANTHIN

A freshwater green microalga, *Haematococcus pluvialis*, has been cultivated as a source of natural astaxanthin, a ketocarotenoid. Astaxanthin in microalgal cells is located in the cytoplasmic lipid globules. The biflagellate and motile cells of

H. pluvialis transform into resting cyst cells, the aplanospores, and develop a distinct red color due to astaxanthin accumulation. After maturation, the cysts germinate, releasing flagellated cells (Margalith, 1999). The transformation of vegetative microalgal cells to astaxanthin-accumulating resting cells could be achieved by subjecting the microalgal culture to environmental and nutritional stress, for example, nitrogen and phosphorus limitation, increases in culture temperature, increases in the salinity of the culture medium, and exposure of the culture to high irradiance (Del Campo et al., 2007). The astaxanthin content of *Haematococcus* cells can go up to 3%, making them an attractive source of the carotenoid pigment.

The accumulation of astaxanthin in *Haematococcus* cells in the resting phase necessitates a two-phase cultivation protocol where in the first phase the microalga is grown under optimal growth conditions to achieve high biomass yields, and then the green biomass is subjected to nutritional and environmental stress in a second phase to induce cyst (aplanospore) formation and the accumulation of astaxanthin (Del Campo et al., 2007). Complete outdoor cultivation of *Haematococcus* has not been feasible due to its high sensitivity to contamination and extreme environmental conditions during the growth phase. Commercial production of *Haematococcus* biomass is generally carried out in closed photobioreactors, or it combines closed photobioreactors and open ponds where the first stage of biomass generation is carried out in closed photobioreactors, followed by a short residence period of culture in open ponds for the second phase of induction of astaxanthin accumulation (Olaizola and Huntley, 2003; Cysewski and Lorenz, 2004; Del Campo, 2007).

Astaxanthin has major commercial application in aquaculture as a source of pigmentation for salmon, trout, and red sea bream (Lorenz and Cysewski, 2000; Guerin et al., 2003; Cysewski and Lorenz, 2004), and the market is dominated by synthetic astaxanthin. Natural astaxanthin from *Haematococcus* is not competitive with synthetic astaxanthin for aquaculture applications due to high production costs (Guerin et al., 2003; Olaizola, 2003). Therefore, the economic viability of large-scale cultivation of *Haematococcus* to produce natural astaxanthin for aquaculture applications alone may not be feasible, but finding high-value markets is important. Human nutraceuticals have emerged as the high-value market for natural astaxanthin from *Haematococcus*. Several in vitro and in vivo studies have demonstrated the beneficial health effects of *Haematococcus*-derived natural astaxanthin (Guerin et al, 2003; Olaizola, 2003; Kamath et al., 2008; Yuan et al., 2011). *Haematococcus* has been cleared by the U.S. FDA for application as an ingredient in dietary supplements for humans and has also been approved for human consumption in several European countries (Lorenz and Cysewski, 2000). This has paved the way for marketing *Haematococcus* biomass for application as a food supplement.

11.4.3 OTHER VALUE-ADDED PRODUCTS (VAPS)

Microalgae have also been recognized as potential sources of various other VAPs for food applications. Green microalgae and cyanobacteria could be used as a rich source of the photosynthetic pigment chlorophyll. Chlorophyll acts as a chelating agent and can be used in ointments, and in the treatment of liver recovery and ulcers (Puotinen, 1999). Chlorophyll can also be used as a natural colorant in foods (Humphrey, 2004). Microalgae that have already been grown in large volumes (e.g., *Chlorella* and *Spirulina*) could be explored for the commercial production of chlorophylls for food application.

Microalgae might also be a potential source for the commercial production of lutein. Lutein usually occurs in microalgae in its free nonesterified form. Microalga *Muriellopsis* sp. has been shown to accumulate higher contents of lutein with high productivities of biomass under photo-autotrophic conditions (Del Campo et al., 2000). Studies on the cultivation of *Muriellopsis* in closed photobioreactors as well as open-pond systems have been carried out (Del Campo et al., 2001, 2007; Blanco et al., 2007). The free lutein content of *Muriellopsis* biomass was found to be in the range of 0.4% to 0.6% on a dry weight basis (Del Campo et al., 2007). *Scenedesmus* sp. and *Chlorella* sp. have also been reported to accumulate lutein (Del Campo et al., 2000, 2004; 2007; Shi et al., 2006).

The red microalga *Porphyridium* has been shown to be a potential source of sulfated polysaccharides that form thermally reversible gels similar to macroalgae-derived polysaccharides, agar, and carrageenan. These gels have various commercial applications, including in foods, as gelling agents, thickeners, stabilizers, and emulsifiers (Raja et al., 2008). As a microalga, *Porphyridium* may offer an advantage over macroalgae due to its relatively faster growth rate. Small-scale outdoor cultivation studies with *Porphyridium* have been carried out (Arad et al., 1985).

The chlorophyll, lutein, and polysaccharides from microalgae could be commercially important VAPs for food applications. However, processes for their commercial production from microalgae have not yet been developed and require further R&D studies as well as the development of markets for these products. Table 11.4 provides a list of some of the microalgal species with relevance for biotechnological applications in food. From Table 11.4, it is very clear that some progress has been made, and there are commercial microalgae applications, including pigments, fatty acids, and health foods. However, studies on the characterization of indigenous microalgae from natural habitats as potential sources of food, feed, and VAPs have remained relatively limited, and there is the need to tap into the vast biodiversity of microalgae growing in natural habitats under diverse climatic conditions for finding suitable candidate microalgae for various applications.

11.5 FUTURE NEEDS

Research efforts into using microalgae for CO_2 sequestration, biodiesel production, and other VAP syntheses will continue to power several of the assets inherent in these photosynthetic organisms. The high lipid content of microalgae has been taken as the major screening criteria for selecting and exploiting such species for biodiesel production, but has not been evaluated critically. The species that have been exploited for biodiesel production are very few (Griffiths and Harrison, 2009). The majority of the research work has focused on increasing lipid content and biomass productivity, whereas the studies related to chemical conversion of lipid to biodiesel, quality improvement, and cost reduction of the process are progressing at a slow pace (Krohn et al., 2011). Taking into consideration the current scenario, there is a need to look into the complete fatty acid profile of microalgal lipids in addition

TABLE 11.4

Biotechnological Application of Some Microalgae Species for Food-Based Applications

Species/Group	Product	Application Areas	Cultivation Systems
Spirulina platensis/ Cyanobacteria	Phycocyanin, biomass	Health food, food color	Open ponds, natural lakes
Chlorella vulgaris/ Chlorophyta	Biomass	Health food, food supplement	Open ponds, basins, glass-tube photobioreactors
Dunaliella salina/ Chlorophyta	Carotenoids, β-carotene	Health food, food supplement	Open ponds, lagoons
*Haematococcus pluvialis/*Chlorophyta	Carotenoids, astaxanthin	Health food, food supplement	Open ponds, closed photobioreactors
Odontella aurita/ Bacillariphyta	Fatty acids	Baby food	Open ponds
*Porphyridium cruentum/*Rhodophyta	Polysaccharides	Nutrition	Tubular photobioreactors
Phaeodactylum tricornutum/ Bacillariophyta	Lipids, fatty acids	Nutrition	Open ponds, basins
Lyngbya majuscola/ Cyanobacteria	Immune modulators	Nutrition	
*Crypthecodinium cohnii/*Dinoflagellata	Docosahexaenoic acid	Supplement in infant formulas, dietary supplement	Heterotrophic fermentation

Source: Adapted from Pulz and Gross (2004); and Raja et al. (2008).

to the qualitative and quantitative profiling of triacylglycerides and free fatty acids (Ramos et al., 2009; Liu et al., 2010). These factors primarily influence the quality of biodiesel produced. Once the right microalgae species have been selected considering all physico-chemical properties, culture conditions can be optimized to obtain higher biomass productivity (Rodolfi et al., 2008) in an economical way in a raceway pond and/or closed photobioreactor system. In addition, an understanding of microalgal behavior at the molecular level during the process of CO_2 tolerance and uptake for intracellular lipid enhancement is a must.

Although CO_2 sequestration by microalgae into biomass and triacylglycerol storage plays a critical role in an organism's ability to withstand stress, information concerning the enzymes of CO_2 uptake and tolerance, triacylglycerol synthesis, their regulation by nutrients, physiological conditions, their mechanisms of action along with the roles of specific isoforms has been limited by the lack of studies on proteomics and genomics (detailed protein and gene profiling) of microalgae for CO_2 sequestration and biodiesel production.

The exploration of the vast biodiversity of microalgae in natural habitats for selection of suitable strains for CO_2 sequestration and VAPs is possible. Potential microalgae tolerating high CO_2 concentrations can be isolated from relevant sources such as lakes, ponds, etc. near thermal power plants. The microalgal strains that can tolerate high CO_2 concentrations and also synthesize food/feed and biofuel precursors need to be developed by exploring the microbial diversity. It should be possible to control the composition of food and biofuel precursors by suitably manipulating stress conditions such as light, temperature, and nutrients. The high performance of cultivation systems (open-pond and/or closed photobioreactor system) for microalgae with high biomass productivity and energy efficiency should possibly be developed through a fundamental understanding of culture behavior as well as gas-liquid mass transfer, reactor hydrodynamics, shear stress profiles, light penetration, photoperiod, etc.

Therefore, future research in this area is required to provide new insights into novel ways to use microalgae in economically viable value-added production processes along with their integration with CO_2 sequestration.

REFERENCES

Abu-Khader, M.M. (2006). Recent progress in CO_2 capture/sequestration: A review. *Energy sources, Part A: Recovery, Utilization, and Environmental Effects*, 28: 1261–1279.

Apt, K.E., and Behrens, P.W. (1999). Commercial developments in microalgal biotechnology. *Journal of Phycology,* 35: 215–226.

Arad, S.M., Adda, M., and Ephraim, C. (1985). The potential of production of sulfated polysaccharides from *Porphyridium. Plant and Soil,* 89: 117–127.

Astorg, P. (1997). Food carotenoids and cancer prevention: An overview of current research. *Trends in Food Science and Technology,* 8: 406–413.

Becker, E.W. (2004). Microalgae in human and animal nutrition. In Richmond, A., Ed. *Handbook of Microalgal Culture. Biotechnology and Applied Phycology.* Oxford: Blackwell Science, pp. 312–351.

Becker, E.W. (1988). Micro-algae for human and animal consumption. In *Micro-Algal Biotechnology.* Borowitzka, M.A., and Borowitzka, L.J. (Eds.). Cambridge: Cambridge University Press, pp. 222–256.

Becker, E.W. (1994). *Microalgae. Biotechnology and Microbiology.* Cambridge: Cambridge University Press, ISBN 978-0-521-06113.

Beer, L.L., Boyd, E.S., Peters, J.W., and Posewitz, M.C. (2009). Engineering algae for biohydrogen and biofuel production. *Current Opinions in Biotechnology,* 20(3): 264–271.

Ben-Amotz, A. (1999). *Dunaliella* β-carotene: From science to commerce. In Seckbach J. (Ed.), *Enigmatic Microorganisms and Life in Extreme Environments.* Kluwer, The Netherlands, pp. 401–410.

Bewicke, D., and Potter, B.A. (1984). *Chlorella the Emerald Food.* Ronin Publishing, Berkeley, CA.

Camacho, R.F., García, C.F., Fernández Sevilla, J.M., Chisti, Y., and Molina Grima, E. (2003). A mechanistic model of photosynthesis in microalgae. *Biotechnology & Bioengineering,* 81: 459–473.

Carlucci, M.J., Scolaro, L.A., and Damonte, E.B. (1999). Inhibitory action of natural carrageenans on *Herpes simplex* virus infection of mouse astrocytes. *Chemotherapy,* 45(6): 429–436.

Chanakya, H.U.N., Mahapatra, D.M., Sarada, R., Chauhan, V.S., and Abitha, R. (2012). Sustainability of large-scale algal biofuel production in India. *Journal of the Indian Institute of Science,* 92(1): 63–98.

Cheng, L., Zhang, L., Chen, H., and Gao, C. (2006). Carbon dioxide removal from air by microalgae cultured in a membrane-photobioreactor. *Separation and Purification Technology,* 50(3): 324–329.

Chisti, Y. (2006). Microalgae as sustainable cell factories. *Environmental Engineering and Management Journal,* 5(3): 261–274.

Chisti, Y. (2007). Biodiesel from microalgae. *Biotechnology Advances,* 25: 294–306.

Chisti, Y. (2008). Biodiesel from microalgae beats bioethanol. *Trends Biotechnology,* 26: 126–131.

Cysewski, G.R., and Lorenz, R.T. (2004). Industrial production of microalgal cell–mass and secondary products-species of high potential. *Haematococcus.* In Richmond, A., Ed. *Handbook of Microalgal Culture. Biotechnology and Applied Phycology.* Oxford: Blackwell Science, pp. 281–288.

Dayananda, C., Kumudha, A., Sarada, R., and Ravishankar, G.A. (2010). Isolation, characterization and outdoor cultivation of green microalgae *Botryococcus* sp. *Scientific Research and Essays,* 5(17): 2497–2505.

Dayananda, C., Sarada, R., Usha Rani, M., Shamala, T.R., and Ravishankar, G.A. (2007). Autotrophic cultivation of *Botryococcus braunii* for the production of hydrocarbons and exopolysaccharides in various media. *Biomass and Bioenergy,* 31: 87–93.

Dayananda, C., Sarada, R., Srinivas, P., Shamala, T.R., and Ravishankar, G.A. (2006). Presence of methyl branched fatty acids and saturated hydrocarbons in botryococcene producing strain of *Botryococcus braunii. Acta Physiologiae Plantarum,* 28(3): 251–256.

Del Campo, J.A., Garcia-Gonzalez, M., and Guerrero, M.G. (2007). Outdoor cultivation of microalgae for carotenoid production: Current state and perspectives. *Applied Microbiology and Biotechnology,* 74: 1163–1174.

Del Campo, J.A., Rodriguez, H., Moreno, J., Varga, M.A., Rivas, J., and Guerrero, M.G. (2001). Lutein production by *Muriellopsis* sp. in an outdoor tubular photobiorector. *Journal of Biotechnology,* 81: 289–295.

Del Campo, J.A., Rodriguez, H., Moreno, J., Varga, M.A., Rivas, J., and Guerrero, M.G. (2000). Carotenoid content of chlorophycean microalgae. Factors determining lutein accumulation in *Muriellopsis* sp. (Chlorophyta). *Journal of Biotechnology,* 76: 51–59.

Del Campo, J.A., Rodriguez, H., Moreno, J., Varga, M.A., Rivas, J., and Guerrero, M.G. (2004). Accumulation of astaxanthin and lutein in *Chlorella zofingiensis* (Chlorophyta). *Applied Microbiology and Biotechnology,* 64: 848–854.

Demming-Adams, B., and Adams, W.W. III (2002). Antioxidants in photosynthesis and human nutrition. *Science,* 298: 2149–2153.

Eriksen, N.T. (2008). Production of phycocyanin–A pigment with applications in biology, biotechnology, foods and medicine. *Applied Microbiology and Biotechnology,* 80: 1–14.

FAO/WHO. Energy and Protein Requirement. Report of a Joint FAO/WHO ad hoc Expert Committee. Vol. 52. Geneva: FAO.

Fargione, J., Hill, J., Tilman, D., Polasky, S., and Hawthorne, P. (2008). Land clearing and the biofuel carbon debit. *Science,* 319: 1235–1238.

Francisco, E.C., Neves, B.D., Lopes, J.E., and Franco, T.T. (2010). Microalgae as feedstock for biodiesel production: Carbon dioxide sequestration, lipid production and biofuel quality. *Journal of Chemical Technology and Biotechnology,* 85: 395–403.

Fulke, A.B., Mudliar, S.N., Yadav, R., Shekh, A., Srinivasan, N., Ramanan, R., Krishnamurthi, K., Devi, S.S., and Chakrabarti, T. (2010). Bio-mitigation of CO_2, calcite formation and simultaneous biodiesel precursor's production using *Chlorella* sp. *Bioresource Technology,* 101: 8473–8476.

Garcia-Gonzalez, M., Moreno, J., Manzano, C., Florencio, F.J., and Guerrero, M.G. (2005). Production of *Dunaliella salina* biomass rich in 9-*cis* β-carotene and lutein in a closed tubular photobioreactor. *Journal of Biotechnology,* 115: 81–90.

Gouveia, L., and Empis, J. (2003). Relative stabilities of microalgal carotenoids in microalgal extracts, biomass and fish feed: Effect of storage conditions. *Innovative Food Science and Emerging Technology,* 4: 227–233.

Griffiths, M.J., and Harrison, T.L. (2009). Lipid productivity as a key characteristic for choosing algal species for biodiesel production. *Journal of Applied Phycology,* 21: 493–507.

Guerin, M., Huntley, M.E., and Olaizola, M. (2003). Haematococcus astaxanthin: Applications for human health and nutrition. *Trends Biotechnology,* 21: 210–216.

Hende, S.V.D., Vervaeren, H., and Boon, N. (2012). Flue gas compounds and microalgae: Bio-chemical interactions leading to biotechnological opportunities. *Bioresource Technology,* 30(6): 1405–1424.

Higuera-Ciapara, I., Felix-Valenzuela, L., and Goycoolea, F.M. (2006). Astaxanthin: A review of its chemistry and applications. *Critical Reviews in Food Science and Nutrition,* 46: 185–196.

Ho, S.H., Chen, C.Y., Yeh, K.L., Chen, W.M., Lin, C.Y., and Chang, J.S. (2010). Characterization of photosynthetic carbon dioxide fixation ability of indigenous *Scenedesmus obliquus* isolates. *Biochemical Engineering Journal,* 53(1): 57–62.

Hu, Q.H. (2004). Industrial production of microalgal cell-mass and secondary products-major industrial species. *Arthrospira (Spirulina) platensis* sp. In Richmond, A., Ed. *Handbook of Microalgal Culture. Biotechnology and Applied Phycology.* Oxford: Blackwell Science, pp. 254–272.

Humphrey, A.M. (2004). Chlorophyll as a colour and functional ingredient. *Journal of Food Science,* 69: 422–425.

Hussein, G., Sankawa, U., Goto, H., Masumoto, K., and Watanabe, H. (2006). Astaxanthin, a carotenoid with potential in human health and nutrition. *Journal of Natural Products,* 69: 443–449.

Intergovernmental Panel on Climate Change [IPCC] (2007). Observed changes in climate and their effect. *Climate Change 2007: Synthesis Report,* pp. 30.

Iwamoto, H. (2004). Industrial production of microalgal cell mass and secondary products-major industrial species. In Richmond, A., Ed. *Handbook of Microalgal Culture. Biotechnology and Applied Phycology,* Oxford: Blackwell Science, pp. 270–281.

Johnson, E., Krinsky, N., and Russel, R. (1996). Serum response of all-*trans* and 9-*cis* isomers of β-carotene in humans. *Journal of the American College of Nutrition,* 15: 620–624.

Kamath, S.B., Srikanta, M., Shylaja, M.D., Sarada, R., and Ravishankar, G.A. (2008). Ulcer preventive and antioxidant properties of astaxanthin from *H. pluvialis.* *European Journal of Pharmacology,* 500: 387–395.

Krinsky, N.I., and Johnson, E.J. (2005). Carotenoid actions and their relation to health and disease. *Molecular Aspects of Medicine,* 26: 459–516.

Kumar, A., Ergas, S., Yuan, X., Sahu, A., Zhang, Q., Dewulf, J., Malcata, F.X., and Langenhove, H.V. (2010). Enhanced CO_2 fixation and biofuel production via microalgae: Recent developments and future directions. *Trends in Biotechnology,* 28: 371–380.

Lam, K.M., and Lee, T.K. (2011). Microalgae biofuels: A critical review of issues, problems and the way forward. *Biotechnology Advances,* doi: 10.1016/j.biotechadv.2011.11.008.

Li, Q., and Du, W., (2008). Perspectives of microbial oils for biodiesel production. *Applied Microbiology and Biotechnology,* 80(5): 749–756.

Liu, J., Huang, J., Fan, K. W., Jiang, Y., Zhong, Y., Sun, Z., and Chen, F. (2010). Production potential of *Chlorella zofingienesis* as a feedstock for biodiesel. *Bioresource Technology,* 101: 8658–8663.

Lorenz, R.T., and Cysewski, G.R. (2000). Commercial potential for *Haematococcus* microalgae as a natural source of astaxanthin. *Trends in Biotechnology,* 18: 160–167.

Margalith, P.Z. (1999). Production of ketocarotenoids by microalgae. *Applied Microbiology and Biotechnology,* 51: 431–438.

Martek Biosciences Corporation. *History of Martek.* Martek Biosciences. http://martek.com/about/history.aspx.

Mata, T.M., Martins, A.A., and Caetano, N.S. (2010). Microalgae for biodiesel production and other applications: A review. *Renewable and Sustainable Energy Reviews,* 14: 217–232.

Metting, F.B. (1996). Biodiversity and application of microalgae. *Journal of Industrial Microbiology,* 17: 477–489.

Molina Grima, E., Acién Fernández, F.G., García Camacho, F., Camacho Rubio, F., and Chisti, Y. (2000). Scale-up of tubular photobioreactors. *Journal of Applied Phycology,* 12: 355–368.

Molini Grima, E., Belarbi, E.H., Fernandez, F.G.A., Medina, A.R., and Chisti, Y. (2003). Recovery of microalgae biomass and metabolites: Process options and economics. *Biotechnology Advances,* 20: 491–515.

National Algal Biofuels Technology Roadmap (2010). U.S. Department of Energy, Office of Energy Efficiency and Renewable Energy, Biomass Program.

Olaizola, M. (2003). Commercial development of microalgal biotechnology: From the test tube to the marketplace. *Biomolecular Engineering,* 20: 459–466.

Olaizola, M., and Huntley, M.E. (2003). Recent advances in commercial production of astaxanthin from microalgae. In Fingerman M., and Nagabhushanam R. (Eds.). *Biomaterials and Bioprocessing.* Enfield Science Publishers. pp. 143–164.

Olaizola, M. (2012). Commercial production of astaxanthin from *Haematococcus pluvialis* using 25,000 litre outdoor photobioreactors. *Journal of Applied Phycology,* 12(3): 499–506.

Pulz, O., and Gross, W. (2004). Valuable products from biotechnology of microalgae. *Applied Microbiology and Biotechnology,* 65: 635–648.

Puotinen, C.J. (1999). *Herbs for Detoxification.* New York: McGraw-Hill Professional, p. 25.

Radmer, R.J. (1996). Algal diversity and commercial algal products. *Bioscience,* 46: 263–270.

Raja, R., Hemaiswarya, S., and Rengasamy, R. (2007). Exploitation of *Dunaliella* for β-carotene production. *Applied Microbiology and Biotechnology,* 74: 517–523.

Raja, R., Hemaiswarya, S., Ashok Kumar, N., Sridhar, S., and Rengaswamy, R. A perspective on biotechnological potential of microalgae. *Critical Reviews in Microbiology,* 34: 34–77.

Ramanan, R., Kannan, K., Devi, S.S., Mudliar, S., Kaur, S., Tripathi, A.K., and Chakrabarti, T. (2009a). Bio-sequestration of carbon dioxide using carbonic anhydrase enzyme purified from *Citrobacter freundii. World Journal of Microbiology Biotechnology,* 25: 981–987.

Ramanan, R., Kannan, K., Vinayagamoorthy, N., Ramkumar, K.M., Devi, S.S., and Chakrabarti, T. (2009b). Purification and characterization of a novel plant-type carbonic anhydrase from *Bacillus subtilis. Biotechnology and Bioprocess Engineering,* 14: 32–37.

Ramos, M.J., Fernández, C.M., Casas, A., Rodriguez, L., and Perez, A. (2009). Influence of fatty acid composition of raw materials on biodiesel properties. *Bioresource Technology,* 100(1): 261–268.

Reifert, H. (2007). Anthropogenic change-CO_2 rising faster. *Proceedings of the National Academy of Sciences U.S.A.,* 104: 10288–10293.

Sanchez, M.A., Contreras, G.A., Garcia, C.F, Molina, G.E., and Chisti, Y. (1999). Comparative evaluation of compact photobioreactors for large-scale monoculture of microalgae. *Journal of Biotechnology,* 70: 249–270.

Sheehan, J., Dunahay, T., Benemann, J., and Roessler, P. (1998). A Look Back at the U.S. Department of Energy's Aquatic Species Program: Biodiesel from Algae. National Renewable Energy Laboratory (NREL), U.S. Department of Energy.

Shekh, A.Y., Krishnamurthi, K., Mudliar, S.N., Yadav, R.R., Fulke, A.B., Devi, S.S., and Chakrabarti, T. (2012). Recent advancements in carbonic anhydrase driven processes for CO_2 sequestration: Mini review. *Critical Reviews in Environmental Science and Technology,* 42(14): 1419–1440.

Shi, X., Zhengyun, W., and Chen, F. (2006). Kinetic model of lutein production by heterotrophic *Chlorella* at various pH and temperature. *Molecular Nutrition and Food Research,* 50: 763–768.

Singh, S., Bhushan, K.N., and Banerjee, U.C. (2005). Bioactive compounds from cyanobacteria and microalgae: An overview. *Critical Review Biotechnology,* 25(3): 73–95.

Spolaore, P., Joannis-Cassan, C., Duran, E., and Isambert, A. (2006). Commercial applications of microalgae. *Journal of Biosciences and Bioengineering,* 10: 87–96; *Energy Science and Research,* 25: 31–43.

Telfer, A. (2002). What is β-carotene doing in the photosystem II reaction centre? *Philosophical Transactions of the Royal Society of London B: Biological Sciences,* 357: 1431–1439.

Verma, N. M., Mehrotra, S., Shukla, A., and Mishra, B. N. (2010). Prospective of biodiesel production utilizing microalgae as the cell factories: A comprehensive discussion. *African Journal of Biotechnology,* 9(10): 1402–1411.

Vilchez, G., Garbayo, I., Lobato, M.V., and Vega, J.M. (1997). Microalgae-mediated chemicals production and wastes removal. *Enzyme and Microbial Technology,* 20: 562–572.

Von Laar, J., Stahl, W., Bolsen, K., Goerz, G., and Sies, H. (1996). β-Carotene serum levels in patients with erythropoietic protoporphyria on treatment with the synthetic all-trans isomer or a natural isomeric mixture of β-carotene. *Journal of Photochemistry and Photobiology: B,* 33: 157–162.

Vorst, P., Baard, R.L., Mur, L.R., Korhals, H.J., and Van, D. (1994). Effect of growth arrest on carotene accumulation photosynthesis in *Dunaliella. Microbiology,* 140: 1411–1417.

Wang, Y., Noguchi, K., and Terashima, I. (2008). Distinct light responses of the adaxial and abaxial stomata in intact leaves of *Helianthus annuus* L. *Plant Cell Environment,* 31: 1307–1316.

Yadav, R.R., Mudliar, S.N., Shekh, A.Y., Fulke, A.B., Devi, S.S., Krishnamurthi, K., Juwarkar, A., and Chakrabarti, T. (2012). Immobilization of carbonic anhydrase in alginate and its influence on transformation of CO_2 to calcite, *Process Biochemistry,* 47(12): 585–590.

Yang, J., Qiu, H.G., Huang, J.K., and Rozelle, S., 2008. Fighting global food price rises in the developing world: The response of China and its effect on domestic and world markets. *Agricultural Economics,* 39, 453–464.

Yeesang, C., and Cheirsilp, B. (2011). Effect of nitrogen, salt, and iron content in the growth medium and light intensity on lipid production by microalgae isolated from freshwater sources in Thailand. *Bioresource Technology,* 102(3): 3034–3040.

Yuan, J.P., Juan, P., Yin, K., and Wang, J.H. (2011). Potential health-promoting effects of astaxanthin: A high-value carotenoid mostly from microalgae. *Molecular Nutrition and Food Research,* 55: 150–165.

Zhao, B., Zhang, Y., Xiong, K., Zhang, Z., Hao, X., and Liu, T. (2011). Effect of cultivation mode on microalgae growth and CO_2 fixation, *Chemical Engineering Research and Design,* 89(9): 1758–1762.

12 Phycoremediation by High-Rate Algal Ponds (HRAPs)

Ismail Rawat, Ramanathan Ranjith Kumar,
and Faizal Bux
Institute for Water and Wastewater Technology
Durban University of Technology
Durban, South Africa

CONTENTS

12.1 INTRODUCTION

The disposal of liquid and solid waste in rivers, streams, lakes, and oceans has been occurring for extended periods of time. Increasing industrialization to serve rapidly expanding urban population needs generates large amounts of wastewater that require treatment before release in order to prevent further environmental deterioration. Point source wastewater contamination has the capacity to "overload" receiving water bodies and is the most widespread threat to environmental water quality. Wastewater generally contains high concentrations of organic and inorganic nutrients, which are among the main causes of irreversible ecological degradation. This disrupts the bio-system

179

and natural recycling processes such as photosynthesis, respiration, nitrogen fixation, evaporation, and precipitation. Effective wastewater treatment and the use of reclaimed wastewater have great potential to help meet fresh water requirements for various domestic and industrial uses, thus somewhat alleviating the need for water in growing urban centers. In industrial and municipal wastewater, reduction of various chemical stacks at sources is not an easy process and is very expensive to treat by conventional treatment methods due to the demand for skilled operators, high capital investment, high operational costs, reliability etc. Complex operation of conventional treatment methods for removing chemicals does not guarantee sludge reduction. Sludge removal is one of the main challenges in sustainable wastewater treatment, but can be accomplished by the Best Available Technique (BAT) to treat the socio-economic aspect of efficient wastewater treatment. This, coupled with potential energy resource recovery, is mandatory and necessary in exploring the feasibility of biological treatment. There has been growing worldwide interest due to decreasing water resources and increasing demand for preservation and the sustainable management of water resources (Garca et al., 2000).

Microalgal cultivation is an attractive biotechnological wastewater treatment method that has potential as an alternative method to conventional treatment. Microalgae are popular bio-resources, as appropriate microalgal technology can add a number of benefits to the treatment process because they have a greater capacity for the treatment of a number of wastewater contaminants. Chinnasamy et al. (2010) observed that a consortium of fifteen native microalgae efficiently reduced more than 96% of carpet mill treated wastewater nutrients within 72 h. Wang et al. (2010) reported rapid decreases in nitrate, phosphate, and metal levels in wastewater treatment over a short period of microalgal cultivation. Microalgal wastewater treatment is an economically viable method of wastewater treatment that has an extensive research history spanning more than 50 years (Oswald et al., 1953; Oswald, 1991; Ruiz et al., 2011). Microalgae-based wastewater removal of nutrients and/or chemicals is achieved by accumulation in, or conversion to, biomass, making it a better biotechnological method for the preservation of freshwater ecosystems (Hoffmann, 1998; Ruiz et al., 2011). Considering that inexpensive effluent can be used as feed for desired microalgal species to produce algae-derived products, while simultaneously removing nutrients, makes it an attractive biological system. Thus, phycoremediation technology is a promising field for applied studies such as in wastewater treatment, and biomass and biofuels production for sustainable energy.

The cultivation of microalgae for wastewater treatment is a high-quality, eco-friendly process with no secondary pollution. Reclaimed effluent produces high-value microalgal metabolites such as lipids, carbohydrates, and proteins. Microalgae are often applied in the tertiary treatment of domestic wastewater in maturation ponds, or in small- to medium-scale municipal wastewater treatment systems (Hanumantha Rao et al., 2011; Rawat et al., 2011). Technologies such as the advanced integrated wastewater pond systems (AIWPS) are commercially available (Oswald, 1991). The most common designs include facultative ponds, which are relatively deep and support surface growth of microalgae. High-rate algal ponds (HRAPs) are a hallmark technology to treat a number of wastewater streams, especially under tropical and subtropical conditions due to the availability of sunlight utilized by microalgae for photosynthesis (Phang et al., 2000; Mustafa et al., 2011). Shallow ponds depend on mechanical mixing for maximum algae production and removal of biological oxygen demand. HRAPs are the most cost-effective reactors for liquid waste management and capture of solar energy, and the capture of atmospheric carbon dioxide, and are used in the treatment of animal wastes (Narkthon, 1996). HRAP wastewater treatment can be highly efficient in reducing bacteria, biological oxygen demand (BOD), and nutrient levels by integrated approaches to recycling wastes. Phycoremediation can be used in the process as a second step after initial anaerobic treatment of high organic wastewater to yield a significant reduction in influent organic matter, such as nitrogen and phosphorus. Harvested microalgae are rich in nutrients such as nitrogen, potassium, and phosphorus, which can be used for animal feed, etc. (Ogbonna et al., 2000; Olguın, 2003; Rawat et al., 2011). Therefore, HRAPs are very appropriate for sanitation in small rural communities because of their simplicity of operation in comparison to conventional technologies such as the activated sludge process. This chapter critically evaluates phycoremediation HRAPs for removal of high organic strength nutrients by means of enriched microalgae.

12.2 WASTEWATER CHARACTERISTICS

Wastewater is one of the main sources of increasing water pollutant levels globally (Gomec, 2010). An understanding of wastewater characteristics is essential in the design and operational processes of wastewater treatment and deserves ample research efforts. Wastewater is divided into two types: (1) municipal wastewater and (2) industrial wastewater. Wastewater is generally a combination of household and industrial waste, depending on the treatment collection system. These are generated during different human activities and are mixed together. Figure 12.1 provides a few

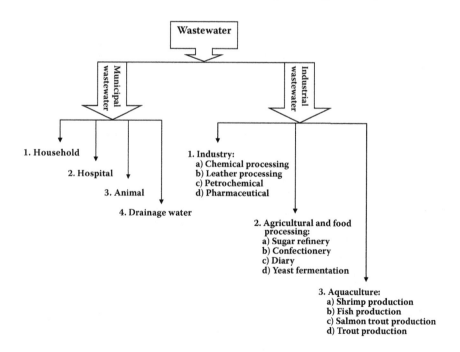

FIGURE 12.1 Types of domestic and industrial wastewater.

TABLE 12.1

Chemical Composition and Characteristics of Untreated Wastewater

Contaminants	Unit	Concentration		
		Weak	Medium	Strong
Total solids	(TS) mg L^{-1}	350	720	1,200
Total dissolved solids	(TDS) mg L^{-1}	250	500	850
Fixed	mg L^{-1}	145	300	525
Volatile	mg L^{-1}	105	200	325
Suspended solids	mg L^{-1}	100	220	350
Fixed	mg L^{-1}	20	55	75
Volatile	mg L^{-1}	80	165	275
Settleable solids	mg L^{-1}	5	10	20
BOD5, 20°C	mg L^{-1}	110	220	400
TOC	mg L^{-1}	80	160	290
COD	mg L^{-1}	250	500	1,000
Nitrogen (total as N)	mg L^{-1}	20	40	85
Organic	mg L^{-1}	8	15	35
Free ammonia	mg L^{-1}	12	25	50
Nitrites	mg L^{-1}	0	0	0
Nitrates	mg L^{-1}	0	0	0
Phosphorus (total as P)	mg L^{-1}	4	8	15
Organic	mg L^{-1}	1	3	5
Inorganic	mg L^{-1}	3	5	10
Chlorides	mg L^{-1}	30	50	100
Sulfate	mg L^{-1}	20	30	50
Alkalinity (as CaCO$_3$)	mg L^{-1}	50	100	200
Grease	mg L^{-1}	50	100	150
Total coliforms	No./100 mL	106–107	107–108	107–109
Volatile organic compounds	g L^{-1}	<100	100–400	>400

Source: From Rawat et al. (2011).

examples of municipal and industrial wastewater types. Municipal and industrial wastewater treatment by socio-economic biological treatment is a highly efficient technology (Metcalf and Eddy, 1991), and reclaimed wastewater can be used for various purposes. Table 12.1 shows the evaluation of weak, medium, and strong typical composition levels of domestic wastewater. A complete assessment of wastewater quality can be broadly classified into three main characteristics by its physical, chemical, and biological constituents and their sources (Figure 12.2).

12.2.1 PHYSICAL CHARACTERISTICS

The perceptual structure of the physical and chemical characteristics of wastewater can change substantially with changes in stream habitats and their individual patterns. Wastewater physical characteristics include

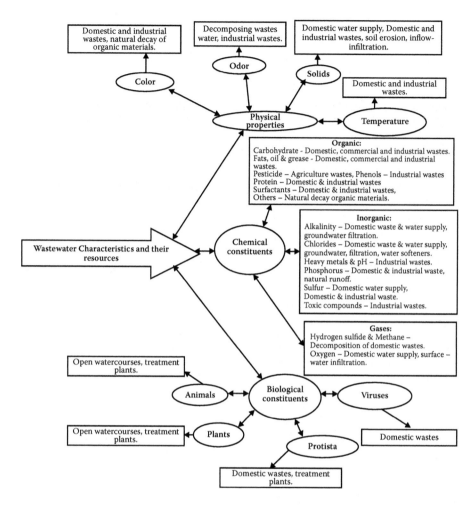

FIGURE 12.2 The physical, chemical, and biological characteristics of wastewater. (*Source*: From Rawat et al., 2011.)

1. *Color:* The physical appearance, qualitative significant color of wastewater depends on holding times in tanks, and varies from a light brown to light gray color. The color is known to turn dark gray or black in the event of wastewater going stale. A color change of wastewater is due to fermentation of the various chemical compounds produced, in particular hydrogen sulfide and ferrous sulfide. Color can be measured by comparisons using standard methods.

2. *Odor:* The offensive odor in wastewater is mainly due to dissolved impurities and a number of odor compounds produced by living enriched microbes and decaying aquatic organisms when under anaerobic conditions. The principal odor-producing compound is hydrogen sulfide, produced as gas by bacterial decomposition under anaerobic conditions.

3. *Solids contents:* The total solids are made up of both dissolved and suspended material that remains as residue in wastewater (Metcalf and Eddy,
1987) upon evaporation at 103°C to 105°C.

4. *Temperature:* The wastewater temperature will vary season to season and
with the geographical location, from 10°C to 21°C (Muttamara, 1996).
Temperature plays a major role in wastewater treatment, and its variation
may cause changes as a result of the chemical and biological reactions of
planktonic organisms. Wastewater contains bacteria and fungi that may
have a substantial influence on the physical characteristics of the wastewater, especially when in abundance due to abnormal temperatures. Turbidity
and color are indirectly related to temperature because most of the chemical
reaction products, including equilibrium of wastewater coagulation, can
change with temperature. Temperature is also very important in the determination of various parameters, such as changes in pH often occurring in
regions with low acid neutralizing capacity, conductivity, different saturation levels of gases, various forms of alkalinity, etc.

12.2.2 CHEMICAL CHARACTERISTICS

The chemical compounds of wastewater characterization are most important with
respect to effective treatment. Identification of the chemical components and
their concentrations are used as a measure of wastewater quality. Domestic and
industrial wastewaters contain a variety of organic and inorganic chemicals. The
principal chemical components in sewage wastewater are carbohydrates, proteins,
lipids, and urea. The urea in wastewater is largely from an organic compound,
urine, which is the chief constituent forming large quantities of nitrogenous matter (Rawat et al., 2011) via rapid decomposition. Organic chemicals, which are
mainly composed of carbon, hydrogen, oxygen, and other components such as
sulfur, phosphorous, iron, ammonia, proteins, fats, lignin, soaps, oils, and other
synthetic organic chemicals that are readily biodegradable and their decomposition products, are found in the system. The physico-chemical parameters in wastewater, such as total dissolved solids (TDS), of the organic chemical characteristics
involve interactions of pH, alkaline minerals, and other nutrients. These are related
to the solvent capabilities of wastewater (Drinan and Whiting, 2001). Some of
the common inorganic chemicals compounds present in wastewater are nitrogen,
sulfur, chloride, phosphorus, irons, hydrogen, and trace amounts of heavy metals
(Muttamara, 1996).

12.2.3 BIOLOGICAL CHARACTERISTICS

Generally in wastewater, millions of microscopic and macroscopic organisms are
widely distributed, originating from discharged domestic wastewater. These include
bacteria, protozoa, viruses, and limited algal species. Many of these micro- and
macroorganisms are considered harmless, and the large diversity of organisms is
highly adapted to their conditions and effective in wastewater treatment and activated sludge treatment within the treatment facility. Several recent publications have

reported that wastewater provides an ideal medium for potential microbial growth (Kong et al., 2010; Cho et al., 2011; Christenson and Sims, 2011; Park et al., 2011a; Pittman et al., 2011; Rawat et al., 2011), irrespective of anaerobic or aerobic wastewater treatment (Abeliovich, 1986).

12.3 PHYCOREMEDIATION

The term *phycoremediation* was coined by John (2000) to refer to the remediation of water carried out by algae. Microalgae have high efficacy in wastewater treatment and can offer possible solutions for environmental problems (Lau et al., 1994; Craggs et al., 1997; Korner and Vermaat, 1998; Harun et al., 2010). Microalgae are eukaryotic, autotrophic microorganisms that can adapt to almost any aquatic environment (including wastewater) and produce biomass rich in various nutrients and minerals. Microalgae vary greatly in protein (10% to 53%), carbohydrate (10% to 16%), lipid (15% to 55%), and mineral (5%) constituents (Xu et al., 2006).

Phycoremediation of wastewater (domestic or industry) refers to any large-scale utilization of (desirable) microalgae for the removal of pollutants or biotransformation of hazardous or harmful organic chemical compounds to nonhazardous end-products, xenobiotics, and removal of pathogens from wastewater. Biomass consumes considerable amounts of nutrients from freely available sources, such as wastewater rich in organic nutrients, inorganic chemicals, and CO_2 from waste and exhaust streams (Olguin, 2003), that can accelerate the microalgal biomass propagation (45% to 60% microalgae by dry weight), nucleic acids, and phospholipids. Nutrient removal can be further increased by ammonia stripping or phosphorus precipitation due to the increase in the pH associated with photosynthesis (Laliberté et al., 1994; Oswald, 2003; Hanumantha Rao et al., 2011; Rawat et al., 2011).

Phycoremediation as a biological tertiary treatment, performed typically to treat secondary municipal wastewater, has been the focus of research during the past few decades (Oswald and Gotaas, 1957). High-rate algal ponds (HRAPs) for wastewater treatment are very effective, in that HRAP-cultivated microalgal cultures can assimilate huge amount of nutrients, resulting in a reduction in BOD and chemical oxygen demand (COD). Microalgae are regarded as the most versatile solution among biological wastewater treatment processes. Domestic wastewater contains the majority of nutrients such as nitrogen and phosphorous that directly and indirectly support microalgal productivity and maintain the biomass at levels high enough to achieve nutrient removal efficiently in wastewater systems. The application of microalgae in wastewater treatment for reducing odor, coloring, nitrate, nitrite, phosphate, ammonia, TDS, TSS, BOD, and increasing pH and heavy metal absorption has been performed over the past few years. Effluent-treated microalgal biomass can be used for various purposes (Munoz and Guieysse, 2006). Recently, Kumar et al. (2011) studied high-rate algal pilot plant cultivated *Chlorella vulgaris* in confectionery effluent wastewater treatment, wherein harvested biomass was used for enzymatic and nonenzymatic antioxidant potential studies. However, the enriched microalgal biomass needs to be harvested at low cost using a cost-effective nutrient removal system. These are still in the infancy stage.

The application and advantages of phycoremediation include (Olguin, 2003)

1. Nutrient removal from both municipal and industrial wastewater or effluent enriched with high organic matter
2. Nutrient and xenobiotic compound removal with the aid of algae-based biosorbents
3. Efficient treatment of acidic and heavy-metal wastewater
4. Increasing oxygenation of the atmosphere
5. CO_2 sequestration
6. Improving effluent quality
7. Transformation and degradation of xenobiotics
8. Biosensing of toxic compounds by algae

12.4   ALGAE SPECIES USED FOR PHYCOREMEDIATION

The long history of research into algae-based wastewater treatment, pioneered by algologists Oswald and co-workers (1953), was designed as a technology to carry out the dual role of microalgae used for wastewater treatment and protein production. It began with Golueke and Oswald (1965), who gained insight into the economic aspects of microalgae-based pond wastewater treatment technology and its potential alternative sources of renovated effluent and protein production. Microalgae have been used extensively as appropriate treatment technologies in pond wastewater treatment since the early 1950s (Oswald et al., 1953; Oswald and Gotaas, 1957; Fallowfield and Garrett, 1985; Lincoln and Earle, 1990; Ghosh, 1991; Oswald, 1991; Borowitzka, 1999; Oswald, 2003; Hanumantha Rao et al., 2011). Phycoremediation can provide a more sustainable long-term solution than any other type of wastewater treatment in which a biological method is employed because microalgae have the greater capacity to fix CO_2 by photosynthesis and efficiently remove nutrients from overloaded wastewaters at minimal cost (Hirata et al., 1996; Murakami and Ikenouchi, 1997). The most efficient nutrient removal from wastewater has been investigated using algal strains with special attributes such as extreme temperature tolerance, chemical composition of high-value by-products, heavy metal accumulation, and mixotrophic growth *inter alia*. The microalgae strain *Phormidium* was isolated from a polar environment below 10°C, and the capability of this strain to remove inorganic nutrients in wastewater during spring and autumn of cold climates was studied by Tang et al. (1997). Common microalgae in wastewater treatment include *Chlorella, Oscillatoria, Scenedesmus, Synechocystis, Lyngbya, Gloeocapsa, Spirulina, Chroococcus, Anabaena,* and others. Among these, the universally grown *Chlorella* species (*vulgaris*) has been used for wastewater treatment throughout the world. They are microalgae that can grow in nitrogen (N) and phosphorous (P) nutrient-enriched municipal wastewater and convert wastewater containing N and P into algal biomass (Green et al., 1995; Benemann and Oswald, 1996; Olguin, 2003; Orpez et al., 2009). Other efficient microalgal species used to remove N and P in various industrial effluents include *Botryococcus braunii, which was* used for primary treated sewage waste (Sawayama et al., 1995); *Scenedesmus obliquus,* which was used in the treatment of urban wastewater (Martinez et al., 2000); and artificial wastewater (Gomez Villa et al., 2005). The pollutants are recovered from the system by harvesting biomass (Adey et al., 1996). Aside from microalgal biomass build-up, luxury reserved materials in the form of pigments, protein, antioxidants, amino acids, and other bioactive compounds make them ideal for stripping nutrients. High-rate wastewater treatment of hazardous or organic pollutants has been carried out by microalgae with special attributes. The most widely studied microalgal strains are *Chlorella*, *Scenedsmus,* and *Ankistrodesmus* species, in which various industry effluents were used, such as paper industry wastewaters, olive oil production waste-water, and mill wastewaters (Ghasemi et al., 2011; Rawat et al., 2011). Microalgal strain selection plays an important role in HRAP wastewater treatment. Microalgal collections house only a few thousand different microalgal strains that can efficiently support wastewater treatment and biomass production for value-added by-products and meet near-future demands for alternate biofuels. Therefore, we need to concentrate on effective microalgal strains in combination with recent advances in genetic engineering and material science to fix the problem.

12.5 HIGH-RATE ALGAL PONDS (HRAPS)

The three general types of maturation ponds employed in wastewater treatment are facultative ponds, anaerobic ponds, and the most common, waste stabilization ponds. Aerobic ponds, also known as high-rate ponds, are shallow and completely oxygen-ated (Oswald, 1978). High-rate algal ponds (HRAPs) were developed beginning in the 1950s as an alternative to unmixed oxidation ponds for BOD, suspended solids, and pathogen removal (Rawat et al., 2011). They constitute a low-cost, low-maintenance technology for the remediation of various types of effluents (De Godos et al., 2010). HRAPs exhibit better performance when compared to anaerobic, aerobic, and facul-tative ponds using the same influent. The co-habitation of photosynthetic algae and heterotrophic bacteria is referred to as HRAP symbiosis. HRAPs have been used for the treatment of a variety of wastewaters, including domestic wastewater, piggery and animal wastewaters, agricultural runoff, and mine drainage and zinc refinery wastewater (Rawat et al., 2011). The utilization of microalgae for the assimilation of nitrogen and phosphorus at low concentrations presents a sustainable alternative to the use of existing treatment systems, as the nitrogen and phosphorus can be recov-ered from the algal biomass for reuse (Boelee et al., 2011). HRAPs are designed to promote algal growth, and the technology generally consists of mechanically mixed shallow raceway ponds (Olguın, 2003; García et al., 2006). A large paddlewheel vane pump is used to create a channel velocity sufficient for gentle mixing. The ponds are generally 2 to 3 m wide, 0.1 to 0.4 m deep, and range from 1,000 to 5,000 m^2 in area, depending on the scale of application (García et al., 2006; De Godos et al., 2009; Rawat et al., 2011). The hydraulic retention time of such systems is generally in the range of 4 to 10 days, depending on climatic conditions. Continuous mixing is pro-vided to keep the cells in suspension and reduce the shading effect, thereby exposing the algae to light periodically, even in denser cultures. The most common design that has proven successful on a large scale is the single-loop paddlewheel mixed. Due to the energy cost dependence on velocity, most ponds have been operated at velocities from 10 to 30 cm s^{-1} (Olguın, 2003; Rawat et al., 2011). The mode of action of the

HRAP occurs directly via growth of algae and harvesting of biomass and indirectly by ammonia-nitrate volatilization and orthophosphate precipitation via a change in pH. Algal photosynthesis thus controls the efficiency of nitrate and phosphate removal (Olguın, 2003). Algal photosynthesis provides oxygen for the decomposition of organic matter by aerobic heterotrophic bacteria, allowing for a reduction in organic matter coupled with the removal of nitrogen and phosphorus due to uptake by the algae (García et al., 2006). The biomass produced as a result can be harvested and used for the production of biofuels via various pathways (Park et al., 2011b).

These systems are simple to operate when compared to conventional technologies, thus making them ideal for use by small rural communities (García et al., 2006). HRAPs have been successfully used in the remediation of piggery effluent and also the effluent from aquaculture systems (Olguın, 2003). The combination of wastewater treatment and biofuel production is receiving much more interest than previously, owing to the advantageous implications of such a combination. However, fundamental large-scale research must be undertaken in order to optimize algal production and maintain high-quality effluent standards (Park et al., 2011a).

12.5.1 NUTRIENT REMOVAL

The removal of nitrogen and phosphorus from wastewater is essential in preventing ecological damage to receiving water bodies. Phosphorus is particularly difficult to remove (Pittman et al., 2011). Chemical precipitation is currently the main commercial process for removing phosphorus from wastewater. Biological removal efficiencies vary from 20% to 30% for most organisms (de-Bashan et al., 2004). The phosphorus is then converted into activated sludge that cannot be fully recycled and is buried in landfills or treated to render sludge fertilizer. Microalgae are effective in removing nitrogen, phosphorus, and toxic metals from wastewater, thus making them ideal candidates for nutrient removal and recovery (Pittman et al., 2011). Microalgal uptake of phosphorus has been shown to be as efficient as chemical treatment (Pittman et al., 2011).

Carbon, nitrogen, phosphorus, and sulfur are essential growth requirements for most microalgae (Chisti, 2007; Tsai et al., 2011; Zeng et al., 2011). These elements are commonly found in domestic wastewater in concentrations that support microalgal cultivation. Minimal nutritional requirements can be estimated using the approximate molecular formula of the microalgal biomass, that is, $CO_{0.48}H_{1.83}N_{0.11}P_{0.01}$ (Chisti, 2007; Putt et al., 2011). Nitrogen is the critical factor for the growth and lipid content regulation of microalgae. Phosphorus, although required in smaller amounts, must be supplied in excess as it complexes with metal ions and is thus not fully bioavailable for cell uptake (Chisti, 2007). Microalgae naturally utilize suitable nutrients and energy sources from their environment, thereby optimizing the efficiency of utilization for growth and survival. They are resilient organisms, in that a single species may be able to undergo various types of metabolism, depending on the available nutrients for growth as well as other environmental factors (Amaro et al., 2011). Nitrogen is utilized in the form of nitrate and ammonia, with ammonia being used preferentially in the presence of both chemical species (Feng et al., 2011).

Phototrophic cultivation uses sunlight and CO_2 as an inorganic carbon source for energy production and growth (Mata et al., 2010). Phototrophic cultivation is less prone to contamination than other types of cultivation. Heterotrophic growth occurs in the absence of light using organic carbon sources such as glucose, acetate, glycerol, fructose, sucrose, lactose, galactose, and mannose (Amaro et al., 2011). Organisms that are able to undergo mixotrophic growth have the ability to photosynthesize or use organic substrates as a carbon source. Mixotrophic production reduces photo-inhibition and decreases the loss of biomass due to dark-phase respiration (Brennan and Owende, 2010; Pittman et al., 2011). Organic carbon sources in wastewater allow microalgae to undergo mixotrophic growth followed by phototrophic growth. This effectively removes nutrients while improving biomass and potential lipid productivity (Feng et al., 2011).

The efficiency of nutrient removal depends on the species of algae cultivated and has been shown to be influenced positively by the cultivation of algal strains that are tolerant to certain extremes, such as extreme temperatures, quick sedimentation, or the ability to grow mixotrophically (Olguin, 2003). Choosing a strain for cultivation in HRAPs should preferentially (1) have a high growth rate, (2) have a high protein concentration when grown under nutrient-limited conditions, (3) be used for animal/fish feed, (4) have the ability to tolerate high nutrient levels, (5) produce a value-added product, (6) be able to grow mixotrophically, and (7) be easily harvested (Sheehan et al., 1998; Olguin, 2003; Rawat et al., 2011). *Chlorella vulgaris, Haematococcus pluvialis, and Arthrospira (Spirulina) platensis,* among others, are examples of species that can grow under photo-autotrophic, heterotrophic, and mixotrophic conditions (Amaro et al., 2011).

Microalgal wastewater treatment has the potential to significantly reduce the costs of treatment when compared to conventional chemical methods; this is partially achieved by negation of the requirement for mechanical aeration as microalgae produce oxygen via the process of photosynthesis (Pittman et al., 2011). The simultaneous treatment of wastewater and production of biomass reduces the cost of both processes (Brennan and Owende, 2010; Christenson and Sims, 2011). Furthermore, the production of biofuel in conjunction with wastewater treatment has been put forward as the most viable method for biofuel production from microalgae in the near future (Brennan and Owende, 2010).

Several studies have proven the potential for nutrient removal from synthetic wastewater by microalgal biomass production. Phosphorus removal of 98% and total ammonia removal has been achieved by (Martinez et al., 2000) using *Scenedesmus obliquus.* Boelee et al. (2011) demonstrated simultaneous removal of nitrate and phosphate to 2.2 mg L^{-1} and 0.15 mg L^{-1}, respectively, using microalgal biofilms. Su et al. (2012) reported phosphorus removal efficiency of algae to be 89%. Certain photosynthetic bacteria and green microalgae such as *Rhodobacter sphaeroides* and *Chlorella sorokiniana* can, under heterotrophic conditions, remove high concentrations of organic acids (>1,000 mg L^{-1}) and ammonia (400 mg L^{-1}) (Olguin, 2003). The bacterial removal of substances such as polycyclic aromatic hydrocarbons, organic solvents, and phenolic compounds may be assisted by the use of microalgae that produce the oxygen required for bacterial action. Heavy-metal biosorption may be achieved by microalgae grown under phototrophic conditions (Brennan and Owende, 2010).

12.5.2 FACTORS AFFECTING HIGH-RATE ALGAE PONDS

The efficiency of HRAPs depends on a variety of factors. Microalgal growth in HRAPs is similar to the production of biomass on artificial media. CO_2, mixing, good light availability and penetration, and essential nutrient content, pH, and temperature are among the most important factors in achieving high biomass production and effective nutrient removal (García et al., 2006; Pittman et al., 2011). Biotic factors such as synergistic bacteria, predatory zooplankton, and pathogenic bacteria may also affect the growth of microalgae. The variables will differ, depending on the type of wastewater and from one wastewater treatment site to another (Pittman et al., 2011). Nutrient content (nitrogen and phosphorus) in wastewater can be significantly higher than in conventional media. Nitrogen in wastewater is generally in the form of ammonia, which can inhibit algal growth at high concentrations (Pittman et al., 2011).

Carbon is assimilated from the atmosphere and CO_2 produced by the oxidation of organic matter. The photosynthetic growth of algae utilizes CO_2 as a carbon source for growth while producing oxygen as a by-product, which is utilized by bacteria to mineralize organic matter and produce CO_2, which is consumed by algal photosynthesis. This aids in the reduction of greenhouse gas emissions (Munoz and Guieysse, 2006; Ansa et al., 2011; Park et al., 2011a). HRAPs are generally carbon limited and must be supplemented, potentially by utilization of flue gas for the improvement of nutrient removal efficiency (De Godos et al., 2010). The diurnal cycle affects photosynthetic activity and thereby pH and nutrient removal efficiency (García et al., 2006). The dissolved CO_2 concentration has a direct effect on the pH of the system, as it is acidic in nature when dissolved in water. The cultivation pH directly affects the bioavailability of nutrients such as ammonia and phosphate. It may also aid in the proliferation of nitrifying bacteria (Craggs, 2005; De Godos et al., 2010). Both the pH and dissolved oxygen (DO) peak at midday due to the maximization of photosynthetic efficiency and thereby the removal of CO_2 and an increase in DO of >200% saturation (García et al., 2006; Park et al., 2011a). The consumption of CO_2 and carbonic acid by photosynthesis increases the pH to basic levels (>11), thereby enhancing nutrient removal via the volatilization of ammonia and phosphorus precipitation (Craggs, 2005; Su et al., 2012). At night, the removal efficiency decreases and may cease due to inadequate oxygen for aerobic respiration. Furthermore, the lower pH at night decreases nitrogen and phosphorus removal due to pH-dependant processes (Garcia et al., 2006; De Godos et al., 2010). High pH may also reduce nutrient utilization via significant inhibition of algal growth due to ammonia toxicity. Furthermore, a pH above 8.3 increasingly inhibits the bacterial activity and thereby the oxidation of organic matter by heterotrophic bacteria (Craggs, 2005; Ansa et al., 2011). The optimal pH for many freshwater algal species is 8, above or below which productivity decreases (Kong et al., 2010). Some algae are, however, capable of growth at pH > 10, such as *Amphora* sp. and *Ankistrodesmus* sp. (Park et al., 2011a). The pH stability in HRAPs is brought about by the balance of CO_2 capture from the air, bacterial respiration, and algal CO_2 uptake (Su et al., 2012).

The productivity of algal cultures—and thus nutrient removal—is light and temperature dependent. Photosynthesis increases with an increase in light intensity until the maximum rate is achieved at light saturation in the absence of nutrient limitation

(Park et al., 2011a). Damage to the light receptors (photo-inhibition) occurs beyond the point of light saturation, thereby reducing productivity (Richmond, 2004). The potential for photo-inhibition to occur is more prevalent during the summer months, resulting in photosynthesis ceasing at midday (Olguin, 2003). With an increase in culture density, there is an increase in the shading effect. An algal concentration of 300 g TSS m^{-3} will absorb all the available light in the top 15 cm of the pond. Mixing is thus essential in reducing this effect (Ansa et al., 2011; Park et al., 2011a). Algal productivity increases with an increase in temperature. For most species of microalgae under optimal culture conditions, optimal temperatures vary between 28°C and 35°C. The optimal temperature varies with nutrient and light limitation. An increase in temperature above the optimal level results in photorespiration, which reduces the overall productivity (Sheehan et al., 1998). Sudden changes in temperature can result in a substantial decline in algal growth. Temperature also affects the pH, oxygen, and CO$_2$ solubility, as well as the ionic equilibrium (Park et al., 2011a).

HRAPs are susceptible to contamination by native algae and grazing by zooplankton and other algal pathogens. Attempts to grow algae as monocultures in HRAPs have failed due to said contamination (Sheehan et al., 1998; Park et al., 2011a). Protozoa and rotifers have the ability to reduce algal concentrations to very low levels in a period of just a few days (Benemann, 2008). Daphnia has the ability to reduce chlorophyll by 99% within a few days. Fungal parasites and viral infections have the ability to induce algal cell structure changes, and changes in diversity and succession, thereby reducing algal populations significantly (Park et al., 2011a; Rawat et al., 2011). Control of grazers and parasites may be achieved by physical methods such as filtration, low DO concentration and high organic loading rates, and chemical treatments such as the application of chemicals that mimic invertebrate hormones, increase the pH, and increase the free ammonia concentration. The most practical method of zooplankton control is the adjustment of the pH to 11, as many zooplanktons have the ability to tolerate low DO levels for extended periods of time. The toxic effects of high pH are augmented by the increase in free ammonia brought about by volatilization of ammonia at high pH. The effects of inhibitory substances on parasitic fungi require elucidation, and no general treatments for fungal control currently exist (Park et al., 2011a).

12.5.3 Efficiency of Wastewater Treatment and Algal Growth

The assimilation of nitrogen and phosphorus into algal and bacterial biomass is seen as advantageous due to the recycling potential of the nutrients via biomass treatment. Unicellular microalgae are found to be the most efficient and most predominant in wastewater treatment ponds (Pittman et al., 2011). The use of combined algae–bacteria cultures increases the nitrogen accumulation efficiency; for example, in the treatment of acetonitrile, 53% ammonia was assimilated into biomass as compared to only 26% in a bacterial system under the same conditions. Under optimal conditions, 100% removal can be achieved (Su et al., 2011). The increased removal efficiency of nutrients may be attributed to the algal requirement of high amounts of nitrogen and phosphorus for the production of proteins, nucleic acids, and phospholipids, which account for 45% to 60% of the algal dry weight (Munoz and Guieysse, 2006). Su et al. (2011) demonstrated COD, ammonia, and phosphate removal efficiencies of up to 98%, 100%, and 72.6%, respectively, for the treatment of municipal wastewater. Nutrient removal efficiencies depend on the cultural conditions as mentioned previously and the nutrient loading rate. Boelee et al. (2011) showed a linear increase in nitrate and phosphate uptake with increasing loading rate up to 1.0 g $m^{-2}d^{-1}$ and 0.13 g $m^{-2}d^{-1}$, respectively, from municipal wastewater. Wang et al. (2011) showed an ammonia removal rate of 90%, irrespective of the initial concentration used. Furthermore, total nitrogen and phosphorus was found to be greatly reduced from piggery wastewater. Nutrient removal efficiencies ranging from 91% to 96% ammonia and 72% to 87% phosphate, depending on the season and depth of the culture, were observed by Olguin (2003).

12.6   WASTEWATER AS FEEDSTOCK FOR BIOMASS PRODUCTION

Microalgal wastewater treatment using microalgae with the production of biomass as a by-product is not a new concept. However, it occurs only on a minor scale in waste stabilization ponds and HRAPs. Wastewater treatment using HRAPs has the potential to produce large amounts of biomass that can be used for a variety of applications, including the production of renewable fuels, fertilizer, animal feed, etc. (Rawat et al., 2011). Recent studies have suggested that the use of wastewater as a substrate for biofuel production may make the process economically viable (Brennan and Owende, 2010; Boelee et al., 2011; Cho et al., 2011). Focusing the growth of microalgae on biomass productivity rather than lipid productivity may be beneficial as larger amounts of biomass improve the viability of conversion to alternate fuels (Pittman et al., 2011). Microalgal biomass to biofuels conversion may be carried out by several methods depending on the biomass characteristics (e.g., lipid or carbohydrate content) (García et al., 2006; Rawat et al., 2011). The yields of biomass from HRAPs depend on the type of effluent being treated with specific regard to nutrient content. Table 12.2 summarizes growth and lipid productivity of microalgal species on a variety of wastewater types. Piggery waste effluent treatment by HRAPs has potential productivities of up to 50 t $ha^{-1}yr^{-1}$ (Rawat et al., 2011).

Maximum algal productivities in HRAPs can be achieved by countering rate-limiting and inhibitory conditions. Carbon is often a rate-limiting substrate and may be alleviated by the addition of CO_2. This addition serves a dual role in the provision of carbon and a method of pH control. The addition of CO_2 has been shown to double algal productivity at the laboratory scale and increase productivity by 30% in a pilot-scale HRAP (Park et al., 2011a). Biomass grown at the Lawrence wastewater treatment plant showed algal productivities ranging from 5 to 16 g $m^{-2}d^{-1}$ and average lipid contents of 10% without the addition of CO_2. With the addition of CO_2, productivities were expected to be 25 g $m^{-2}d^{-1}$ (Sturm and Lamer, 2011). However, it must be considered that addition of excess CO_2 leads to a decrease in pH. A pH maintained at a maximum of 8 inhibits physico-chemical processes of nutrient removal such as volatilization of ammonia and phosphate precipitation (Craggs, 2005). But this is not necessarily a negative point, as the increase in assimilation by biomass production offsets the losses on physico-chemical removal. Furthermore, it enables the recycling of nutrients that would have been otherwise lost. Ammonia volatilization accounts for approximately 24% nitrogen loss in HRAPs without pH control (Park et al., 2011a).

TABLE 12.2

Biomass and Lipid Productivities of Microalgae Grown on Various Wastewater Streams

Wastewater Type	Microalgal Species	Biomass (DW) Productivity (mg L^{-1}d^{-1})	Lipid Content (%DW)	Lipid Productivity (mg L^{-1}d^{-1})
Municipal (primary treated)	nd	25[a]	nd	nd
Municipal (centrate)	*Chlamydomonas reinhardtii* (biocoil grown)	2000	25.25	505
Municipal (secondary treated)	*Scenedesmus obliquus*	26[b]	31.4[i]	8[i]
Municipal (secondary treated)	*Botryococcus braunii*	345.6[c]	17.85	62
Municipal (primary treated + CO$_2$)	*Mix of Chlorella* sp., *Micractinium* sp., *Actinastrum* sp.	270.7[d]	9	24.4
Agricultural (piggery manure with high NO$_3$-N)	*Botryococcus braunii*	700[e]	nd	69
Agricultural (dairy manure with polystyrene foam support)	*Chlorella* sp.	2.6 g m^{-2}d^{-1}	9[i]	230[i] mg m^{-2}d^{-1}
Agricultural (fermented swine urine)	*Scenedesmus* sp.	6[f]	0.9[i]	0.54[i]
Agricultural (anaerobically digested diary manure)	*Mix of Microspora willeana, Ulothrix zonata, Ulothrix aequalis, Rhizoclonium hieroglyphicum, Oedogonium* sp.	5.5 g m^{-2}d^{-1}	nd	nd
Agricultural (swine effluent, maximum manure loading rate)	*R. hieroglyphicum*	10.7 g m^{-2}d^{-1}	0.7[i]	72[i] mg m^{-2}d^{-1}
Agricultural (swine effluent, +CO$_2$, maximum manure loading rate)	*R. hieroglyphicum*	17.9	1.2[i]	210 mg m^{-2}d^{-1}
Agricultural (digested dairy manure, 20× dilution	*Chlorella* sp.	81.4[g]	13.6[i]	11[i]

(Continued)

TABLE 12.2 (*Continued*)
Biomass and Lipid Productivities of Microalgae Grown on Various Wastewater Streams

Wastewater Type	Microalgal Species	Biomass (DW) Productivity (mg $L^{-1}d^{-1}$)	Lipid Content (%DW)	Lipid Productivity (mg $L^{-1}d^{-1}$)
Agricultural (dairy wastewater, 25% dilution)	*Mix of Chlorella* sp., *Micractinium* sp., *Actinastrum* sp.	59[h]	29	17
Industrial (carpet mill, untreated)	*Botryococcus braunii*	34	13.2	4.5
Industrial (carpet mill, untreated)	*Chorella saccharophila*	23	18.1	4.2
Industrial (carpet mill, untreated)	*Dunaliella tertiolecta*	28	15.2	4.3
Industrial (carpet mill, untreated)	*Pleurochrysis carterae*	33	12	4
Artificial wastewater	*Scenedesmus* sp.	126.54	12.8	16.2

Source:　From Rawat et al. (2011).

Note:　nd – Not determined.

[a]　Estimated from biomass value of −1000 mg L^{-1} after 40 days.
[b]　Estimated from biomass value of 1.1 mg L^{-1} h^{-1}.
[c]　Estimated from biomass value of 14.4 mg L^{-1} h^{-1}.
[d]　Estimated from biomass value of 812 mg L^{-1} after 3 days.
[e]　Estimated from biomass value of 7 g L^{-1} after 10 days.
[f]　Estimated from biomass value of 197 mg L^{-1} after 31 days.
[g]　Estimated from biomass value of 1.71 g L^{-1} after 21 days.
[h]　Estimated from lipid productivity and lipid content value.
[i]　Fatty acid content and productivity determined rather than total lipid.

12.7　ECONOMICS AND ENERGY BALANCE OF PHYCOREMEDIATION USING HRAPS

Some researchers have suggested that the production of algal biodiesel would produce greenhouse emissions, increase the water footprint, and require more energy than the production of biofuels from corn and canola feedstocks (Park et al., 2011a). These could, however, offset the use of wastewater as a nutrient source. Other researchers concluded that algal biodiesel production would be energetically viable but feedstock inputs would account for almost half of the energy produced, thus making the process not economically viable. The use of wastewater will offset this, giving a net increase in the net energy ratio (NER) (Sturm and Lamer, 2011). Conversion of algal biomass to energy via a multistage biorefinery process, including lipid extraction for biodiesel, utilization of residual biomass for combustion, and anaerobic digestion of biosolids, has the potential to provide a significant amount of energy in the region of 4,610 kW-h d^{-1} to 48,000 kW-h d^{-1} (Sheehan et al., 1998; Sturm and Lamer, 2011). The energy requirement of conventional wastewater treatment is significantly higher than that of high-rate algal ponds. The Advanced Integrated Wastewater Ponds System (AIWPS), designed by Oswald and Green, LLC, requires up to 91% less energy (kW-h kg^{-1} BOD removed) than conventional systems (Olguin, 2003; Rawat et al., 2011).

Microalgal oxygen release provides the oxygen required for the proliferation of heterotrophic bacteria, thus negating the requirement for mechanical aeration as in conventional wastewater treatment. Conventional wastewater treatment costs approximately four times more than the use of HRAPs (Rawat et al., 2011). The AIWPS consists of advanced facultative ponds with anaerobic digestion pits, HRAPs, algal settling ponds, and maturation ponds in series (Craggs, 2005). This system requires 50 times more land area than conventional wastewater treatment viz. activated sludge, not taking into account the land area required for waste activated sludge disposal. Capital costs and operational costs of the AIWPS are half and less than one-fifth that of activated sludge, respectively (Park et al., 2011a). The supply of nutrients, water, and CO_2 contributes from 10% to 30% of the total cost of commercial algal production (Benemann, 2008). Much of the cost of wastewater HRAPs is covered by the cost of wastewater treatment (Table 12.2). The costs of algal production and harvesting using wastewater treatment HRAPs have less environmental impact in terms of water footprint, energy, and fertilizer use. Recycling of growth media is used as a method of minimizing costs. Recycling can, however, cause a reduction in algal productivity due to the increase in contamination and/or the accumulation of inhibitory metabolites (Park et al., 2011a).

12.8 CONCLUSION

Researchers are in general agreement that the use of wastewater treatment HRAPs is the only economical method currently available for algal production of biofuels. There are significant benefits to the use of wastewater HRAPs for the effective, low-cost treatment of wastewaters and algal biomass production for biofuels generation. There is, however, still a great need to optimize conditions for algal growth and nutrient removal under prevailing climatic conditions. Large-scale lipid optimization and harvesting of algal biomass still remains a challenge, and improvements in this area will subsequently decrease the overall cost of algae production and remediation of wastewater.

ACKNOWLEDGMENTS

The authors hereby acknowledge the National Research Foundation (South Africa) for financial contribution.

REFERENCES

Abeliovich, A. (1986). Handbook of microbial mass culture. In *Algae in Wastewater Oxidation Ponds* (Ed. A. Richmond), CRC Press, Boca Raton, FL, pp. 331–338.

Adey, W.H., Luckett, C., and Smith, M. (1996). Purification of industrially contaminated groundwaters using controlled ecosystems. *Ecological Engineering,* 7: 191–212.

Amaro, H.M., Guedes, A.C., and Malcata, F.X. (2011). Advances and perspectives in using microalgae to produce biodiesel. *Applied Energy,* 88, 3402–3410.

Ansa, E.D.O., Lubberding, H.J., Ampofo, J.A., and Gijzen, H.J. (2011). The role of algae in the removal of *Escherichia coli* in a tropical eutrophic lake. *Ecological Engineering,* 37: 317–324.

Benemann, J.R., and Oswald, W.J. (1996). *Systems and Economic Analysis of Microalgae Ponds for Conversion of Carbon Dioxide to Biomass.* Pittsburgh Energy Technology Center, Pittsburgh, PA, p. 201.

Benemann, J.R. (2008). Overview: Algae oil to biofuels. In *NREL-AFOSR Workshop, Algal Oil for Jet Fuel Production.* Benemann Associates, Arlington, VA.

Boelee, N.C., Temmink, H., Janssen, M., Buisman, C.J.N., and Wijffels, R.H. (2011). Nitrogen and phosphorus removal from municipal wastewater effluent using microalgal biofilms. *Water Research,* 45: 5925–5933.

Borowitzka, M.A. (1999). Commercial production of microalgae: Ponds, tanks, tubes and fermenters. *Journal of Biotechnology,* 70: 313–321.

Brennan, L., and Owende, P. (2010). Biofuels from microalgae—A review of technologies for production, processing, and extractions of biofuels and co-products. *Renewable and Sustainable Energy Reviews,* 14: 557–577.

Chisti, Y. (2007). Biodiesel from microalgae. *Biotechnology Advances,* 25: 249–306.

Chinnasamy, S., Bhatnagar, A., Hunt, R.W., and Das, K.C. (2010). Microalgae cultivation in a wastewater dominated by carpet mill effluents for biofuel applications. *Bioresource Technology,* 101: 3097–3105.

Cho, S., Luong, T.T., Lee, D., Oh, Y.K., and Lee, T. (2011). Reuse of effluent water from a municipal wastewater treatment plant in microalgae cultivation for biofuel production. *Bioresource Technology,* 102: 8639–8645.

Christenson, L., and Sims, R. (2011). Production and harvesting of microalgae for wastewater treatment, biofuels, and bioproducts. *Biotechnology Advances.* doi: 10.1016/j.biotechadv.2011.05.015.

Craggs, R.J., McAuley, P.J., and Smith, V.J. (1997). Wastewater nutrient removal by marine microalgae grown on a corrugated raceway. *Water Research,* 31: 1701–1707.

Craggs, R.J. (2005). Advanced integrated wastewater ponds. In *Pond Treatment Technology, IWA Scientific and Technical Report Series* (Ed. A. Shilton). IWA, London, pp. 282–310.

De-Bashan, L.E., Hernandez, J.P., Morey, T., and Bashan, Y. (2004). Microalgae growth-promoting bacteria as "helpers" for microalgae: A novel approach for removing ammonium and phosphorus from municipal wastewater. *Water Research,* 38: 466–474.

De Godos, I., Blanco, S., García-Encina, P.A., Becares, E., and Muñoz, R. (2009). Long-term operation of high rate algal ponds for the bioremediation of piggery wastewaters at high loading rates. *Bioresource Technology,* 100: 4332–4339.

De Godos, I., Blanco, S., García-Encina, P.A., Becares, E., and Muñoz, R. (2010). Influence of flue gas sparging on the performance of high rate algae ponds treating agro-industrial wastewaters. *Journal of Hazardous Materials,* 179: 1049–1054.

Drinan, J.E., and Whiting, N.E. (2001). *Water & Wastewater Treatment: A Guide for the Nonengineering Professional.* CRC Press, Technomic Publishing, Boca Raton, FL.

Fallowfield, H.J., and Garrett, M.K. (1985). The treatment of wastes by algal culture. *Journal of Applied Bacteriology—Symposium Supplement,* 187S–205S.

Feng, Y., Li, C., and Zhang, D. (2011). Lipid production of *Chlorella vulgaris* cultured in artificial wastewater medium. *Bioresource Technology,* 102: 101–105.

Garca, J., Mujeriego, R., and Hernandez-Marine, M. (2000). High rate algal pond operating strategies for urban wastewater nitrogen removal. *Journal of Applied Phycology,* 12: 331–339.

García, J., Green, B.F., Lundquist, T., Mujeriego, R., Hernández-Mariné, M., and Oswald, W.J. (2006). Long term diurnal variations in contaminant removal in high rate ponds treating urban wastewater. *Bioresource Technology,* 97: 1709–1715.

Ghasemi, Y., Rasoul-Amini, S., and Fotooh-Abadi, E. (2011) The biotransformation, bio-diegradation, and bioremediation of organic compounds by microalgae. *Journal of Phycology,* 47: 969–980.

Ghosh, D. (1991). Ecosystem approach to low-cost sanitation in India. In *Ecological Engineerin for Wastewater Treatment* (Eds. C. Etnier and B. Guterstam). Bokeskogen, Gothenburg, Sweden, pp. 63–79.

Golueke, C.G., and Oswald, W.J. (1965). Harvesting and processing sewage-grown planktonic algae. *Journal of Water Pollution Control Federation,* 37: 471–498.

Gomec, C.Y. (2010). High-rate anaerobic treatment of domestic wastewater at ambient operating temperatures: A review on benefits and drawbacks. *Journal of Environmental Science and Health, Part A, Toxic/Hazardous Substances and Environmental Engineering,* 45: 1169–1184.

Gomez Villa, H., Voltolina, D., Nieves, M., and Pina, P. (2005). Biomass production and nutrient budget in outdoor cultures of *Scenedesmus obliquus* (Chlorophyceae) in artificial wastewater, under the winter and summer conditions of Mazatlan, Sinaloa, Mexico. *Vie et Milieu,* 55: 121–126.

Green, F.B., Lundquist, T.J., and Oswald, W.J. (1995). Energetics of advanced integrated wastewater pond systems. *Water Science and Technology,* 31: 9–20.

Hanumantha Rao, P., Ranjith Kumar, R., Raghavan, B.G., Subramanian, V.V., and Sivasubramanian, V. (2011). Application of phycoremediation technology in the treatment of wastewater from a leather-processing chemical manufacturing facility. *Water South Africa,* 37: 7–14.

Harun, R., Singh, M., Forde, G.M., and Danquah, M.K. (2010). Bioprocess engineering of microalgae to produce a variety of consumer products. *Renewable and Sustainable Energy Reviews,* 14: 1037–1047.

Hirata, S., Hayashitani, M., Taya, M., and Tone, S. (1996). Carbon dioxide fixation in batch culture of *Chlorella* sp. using a photobioreactor with a sunlight-collection device. *Journal of Fermentation Bioengineering,* 81: 470–472.

Hoffmann, J. (1998). Wastewater treatment with suspended and nonsuspended algae *Journal of Phycology,* 34: 757–763.

John, J. (2000). A self-sustainable remediation system for acidic mine voids. In *4th International Conference of Diffuse Pollution,* p. 506–511.

Kong, Q.X., Li, L., Martinez, B., Chen, P., and Ruan, R. (2010). Culture of microalgae *Chlamydomonas reinhardtii* in wastewater for biomass feedstock production. *Applied Biochemistry and Biotechnology,* 160: 9–18.

Korner, S., and Vermaat, J.E. (1998). The relative importance of *Lemna gibba* L., bacteria and algae for the nitrogen and phosphorus removal in duckweed-covered domestic wastewater. *Water Research,* 32: 3651–3661.

Kumar, R.R., Rao, P.H., Subramanian, V.V., and Sivasubramanian, V. (2011). Enzymatic and non-enzymatic antioxidant potentials of *Chlorella vulgaris* grown in effluent of a confectionery industry. *Journal of Food Science and Technology,* doi 10.1007/s13197-011-0501-2.

Laliberté, G., Proulx, G., Pauw, N., and De la Noüe, J. (1994). Algal technology in wastewater treatment. *Ergebnisse der Limnologie,* 42: 283–302.

Lau, P.S., Tam, N.F.Y., and Wong, Y.S. (1994). Influence of organic-N sources on an algal wastewater treatment system. *Resource Conservation and Recycling,* 11: 197–208.

Lincoln, E.P., and Earle, J.F.K. (1990). Wastewater treatment with microalgae. In *Introduction to Applied Phycology* (Ed. I. Akatsuka). SPB Academic Publ., The Hague, Netherlands, pp. 429–446.

Martinez, M.E., Sanchez, S., Jimenez, J.M., El Yousfi, F., and Munoz, L. (2000). Nitrogen and phosphorus removal from urban wastewater by the microalga *Scenedesmus obliquus. Bioresource Technology,* 73: 263–272.

Mata, T.M., Martins, A.A., and Caetano, N.S. (2010). Microalgae for biodiesel production and other applications: A review. *Renewable and Sustainable Energy Reviews,* 14: 217–232.

Metcalf, E., and Eddy, H. (1987). *Wastewater Engineering: Treatment, Disposal, Reuse.* Tata McGraw-Hill Publishing Company Ltd., New York.

Metcalf, E., and Eddy, H. (1991). *Wastewater Engineering: Treatment, Disposal, Reuse.* Tata McGraw-Hill Publishing Company Ltd., New York.

Munoz, R., and Guieysse, B. (2006). Algal-bacterial processes for the treatment of hazardous contaminants: A review. *Water Research,* 40: 2799–2815.

Murakami, M., and Ikenouchi, M. (1997). The biological CO_2 fixation and utilization project by RITE (2). *Energy Conversion and Management,* 38: S493–S497.

Mustafa, E.-M., Phang, S.-M., and Chu, W.-L. (2011). Use of an algal consortium of five algae in the treatment of landfill leachate using the high-rate algal pond system. *Journal of Applied Phycology,* doi: 10.1007/s10811-011-9716-x.

Muttamara, S. (1996). Wastewater characteristics. *Resources, Conservation and Recycling,* 16: 145–159.

Narkthon, S. (1996). Nitrogen and phosphorus removal from piggery wastewater by green algae *Chlorella vulgaris.* In *Faculty of Graduated Studies.* Mahidol University, Bangkok, Thailand.

Orpez, R., Martinez, M. E., Hodaifa, G., El, Y.F., Jbari, N., and Sanchez, S. (2009). Growth of the microalga *Botryococcus braunii* in secondarily treated sewage. *Desalination,* 246: 625–630.

Ogbonna, J.C., Yoshizawa, H., and Tanaka, H. (2000). Treatment of high strength organic wastewater by a mixed culture of photosynthetic microorganisms. *Journal of Applied Phycology,* 12: 277–284.

Olguin, E.J. (2003). Phycoremediation key issues for cost-effective nutrient removal processes. *Biotechnology Advances,* 22: 81–91.

Oswald, W., and Gotaas, H. (1957). Photosynthesis in sewage treatment. *Transactions of the American Society for Civil Engineering,* 122: 73–105.

Oswald, W. (2003). My sixty years in applied algology. *Journal of Applied Phycology,* 15: 99–106.

Oswald, W.J., Gotaas, H.B., Ludwig, H.F., and Lynch, V. (1953). Algal symbiosis in oxidation ponds. *Sewage and industrial Waste,* 25: 692–704.

Oswald, W.J. (1978). *Engineering Aspects of Microalgae. Handbook of Microbiology.* CRC Press, Boca Raton, FL. pp. 519–552.

Oswald, W.J. (1991). Introduction to advanced integrated wastewater ponding systems. *Water Science Technology,* 24: 1–7.

Park, J.B., Craggs, R.J., and Shilton, A.N. (2011a). Wastewater treatment high rate algal ponds for biofuel production. *Bioresource Technology,* 10: 35–42.

Park, J.B., Craggs, R.J., and Shilton, A.N. (2011b). Recycling algae to improve species control and harvest efficiency from a high rate algal pond. *Water Reserach,* 45: 6637–6649.

Phang, S.M., Miah, M.S., Yeoh, B.G., and Hashim, M.A. (2000). *Spirulina* cultivation in digested sago starch factory wastewater. *Journal of Applied Phycology,* 12: 395–400.

Pittman, J.K., Dean, A.P., and Osundeko, O. (2011). The potential of sustainable algal biofuel production using wastewater resources. *Bioresource Technology,* 102: 17–25.

Putt, R., Singh, M., Chinnasamy, S., and Das, K.C. (2011). An efficient system for carbonation of high-rate algae pond water to enhance CO_2 mass transfer. *Bioresource Technology*, 102: 3240–3245.

Rawat, I., Ranjith Kumar, R., Mutanda, T., and Bux, F. (2011). Dual role of microalgae: Phycoremediation of domestic wastewater and biomass production for sustainable biofuels production. *Applied Energy*, 88: 3411–3424.

Richmond, A. (2004). Principles for attaining maximal microalgal productivity in photobioreactors: An overview. *Hydrobiologia*, 512: 33–37.

Ruiz, J., Alvarez, P., Arbib, Z., Garrido, C., Barragán, J., and Perales, J.A. (2011). Effect of nitrogen and phosphorus concentration on their removal kinetics in treated urban wastewater by *Chlorella vulgaris*. *International Journal of Phytoremediation*, 13: 884–896.

Sawayama, S., Inoue, S., Dote, Y., and Yokoyama, S.Y. (1995). CO_2 fixation and oil production through microalgae. *Energy Conversion and Management*, 36: 729–731.

Sheehan, J., Dunahay, T., Benemann, J., and Roessler, P. (1998). A Look Back at the U.S. Department of Energy's Aquatic Species Program—Biodiesel from Algae: Report NREL/TP-580-24190. National Renewable Energy Laboratory, Golden, CO.

Sturm, B.S.M., and Lamer, S.L. (2011). An energy evaluation of coupling nutrient removal from wastewater with algal biomass production. *Applied Energy*, doi:10.1016/j.apenergy.2010.1012.1056.

Su, Y., Mennerich, A., and Urban, B. (2011). Municipal wastewater treatment and biomass accumulation with a wastewater-born and settleable algal-bacterial culture. *Water Research*, 45: 3351–3358.

Su, Y., Mennerich, A., and Urban, B. (2012). Synergistic cooperation between wastewater-born algae and activated sludge for wastewater treatment: Influence of algae and sludge inoculation ratios. *Bioresource Technology*, 105: 67–73.

Tang, E.P.Y., Vincent, W.F., Proulx, D., Lessard, P., and Noue, J. (1997). Polar cyanobacteria versus green algae for tertiary waste-water treatment in cool climates. *Journal of Applied Phycology*, 9: 371–381.

Tsai, D.D.-W., Ramaraj, R., and Chen, P.H. (2011). Growth condition study of algae function in ecosystem for CO_2 bio-fixation. *Journal of Photochemistry and Photobiology B: Biology*, pp. 1011–1344.

Wang, H., Xiong, H., Hui, Z., and Zeng, X. (2011). Mixotrophic cultivation of *Chlorella pyrenoidosa* with diluted primary piggery wastewater to produce lipids. *Bioresource Technology*, 104: 215–220.

Wang, L., Min, M., Li, Y., Chen, P., Chen, Y., Liu, Y., Wang, Y., and Ruan, R. (2010). Cultivation of green algae *Chlorella* sp. in different wastewaters from municipal wastewater treatment plant. *Applied Biochemistry and Biotechnology*, 162: 1174–1186.

Xu, H., Miao, X., and Wu, Q. (2006). High quality biodiesel production from a microalga *Chlorella* protothecoides by heterotrophic growth in fermenters. *Journal of Biotechnology*, 126: 499–507.

Zeng, X., Danquah, M.K., Chen, X. D., and Lu, Y. (2011). Microalgae bioengineering: From CO_2 fixation to biofuel production. *Renewable and Sustainable Energy Reviews*, 15: 3252–3260.

13 Microalgal Biotechnology: Today's (Green) Gold Rush

Ravi V. Durvasula and Durvasula V. Subba Rao
Center for Global Health, Department of Internal Medicine
University of New Mexico School of Medicine and
The Raymond G. Murphy VA Medical Center
Albuquerque, New Mexico, USA

Vadrevu S. Rao
Department of Mathematics
Jawaharlal Nehru Technological University Hyderabad
Kukatpally Campus, Hyderabad, India

CONTENTS

13.1 INTRODUCTION

As the global use of energy is projected to increase fivefold by 2100, several countries are investing in microalgal biotechnology as a source of renewable energy to enhance their energy security. Although microalgae are a source of high-value chemicals such as nutraceuticals and pharmaceuticals, with a gold-rush mentality many entrepreneurs focus primarily on biofuel as an end-product utilizing a few selected "traditional" algal species not native to the region, and extrapolate results obtained from controlled laboratory culture to large-scale outdoor production systems. For optimization of harvesting algal biomass, it would be crucial to know that wide intra- and interspecific variations in the biochemical constituents of microalgae exist, depending on their growth conditions. For example, in eight algal species, the percent lipid per dry weight ranged from 5 to 63, lipid production 10.3 to 90 mg $L^{-1}d^{-1}$, biomass 0.003 to 2.5 g $L^{-1}d^{-1}$, and biomass production on an areal basis from 0.91 to 38 g $m^{-2}d^{-1}$. Also, the commercially important carotene content in *Dunaliella* strain B32 and strain I3 isolated from the Bay of Bengal varied from 0.68 pg carotene per cell to 17.54 pg carotene per cell. As microalgae are renewable, sustainable, and affordable, their potential to produce biofuels and bioactive compounds is great. However, we argue that (1) improvements in strain selection, particularly the extremophile microalgae that have the required properties for large-scale biotechnology; (2) biochemical modification; (3) utility of engineered "designer algal strains"; (4) optimization of growth, biomass production, and harvesting; and (5) enhancement of extraction of biofuel and conversion to co-products would all be necessary to make microalgal biotechnology an economically viable enterprise. A robust bio-economy built on a platform of innovative microalgal technologies is recommended.

Photosynthetic microalgae have been cultivated (Miquel, 1893) and utilized to support the production of animal life in the sea (Allen and Nelson, 1910). The most common "traditional" species used for biotechnology, usually isolated from temperate waters, include *Botryococcus braunii, Chaetoceros calcitrans, Chlamydomonas reinhardtii, Chlorella vulgaris, Chroomonas* sp., *Dunaliella bardawil, D. salina, D. tertiolecta, Haematococcus pluvialis, Isochrysis galbana, Nannochloropsis oculata, Neochloris oleoabundans, Phaeodactylum tricornutum, Rhodomonas* sp., *Scenedesmus obliquus, Skeletonema costatum, Spirulina maxima*, and *Tetraselmis chuii*. Usual practice involves the purchase of a few "traditional" species from a culture center for large-scale propagation, although quite a few researchers are looking at isolating species adapted to local environments.

In addition to utilizing algae as biofeed, there is a global surge in microalgal biotechnology activities for commercial applications such as biofuel, bioactive compounds, and bioremediation. From virtually none in 1990, the total number of publications on microalgal biotechnology leapt to 153 by June 2011; of these, 103 were on microalgal biofuel. This surge coincides with the 1991 Gulf War, when the mind-set of several countries changed to reduce their dependence on imported crude oil and to enhance their energy security. The annual worldwide consumption of motor fuel is 320 billion gallons, of which United States accounted for 44% (http://eia.doe.gov/pub/internationjal/iea 2005/table35.xls). At the current rate of usage, the global use of energy will increase fivefold by 2100 (Huesmann, 2000), prompting major investments in renewable energy. Since 2007, the United States alone has injected more than $1 billion into algae-to-energy research and development.

Microalgal biotechnology has received global attention and the attributive advantages include (1) cultivability on nonarable land, (2) bioremediation of wastewater by growing photosynthetic algal biomass, (3) ease of access to metabolic products that are stored intracellularly, (4) production of biofuel and value-added co-products, and (5) carbon sequestration, a result of the accelerated growth of microalgae for biofuel production. Photosynthetic production of algal biomass can be enhanced by an extraneous enrichment with CO_2; industrial effluents containing CO_2 can be utilized to sustain high algal productivity (Raven, 2009; Benemann, 1993). This could help a nation lower its emissions of greenhouse gases and could be used for carbon tax credit. The International Energy Agency (IEA) estimated that biofuels contribute to approximately 2% of global transport fuel today but could increase to 27% by the year 2050. They project that if biofuel production is sustained, it could displace enough petroleum to avoid the equivalent of 2.1 Gt y^{-1} CO_2 emission—comparable to the net CO_2 absorbed by the oceans calculated by Fairley (2011).

The algae-to-biotechnology framework has five stages—that is, algal cultivation, biomass harvesting, algal oil extraction, oil residue conversion, and by-product distribution—and each has several composite processes (Natural Resources Defense Council, 2009). Given the vast potential of microalgal biotechnology, many entrepreneurs focus largely on algal biomass as a source of biofuel rather than high-value chemicals such as nutraceuticals and pharmaceuticals. For example, by the end of this decade, the projected worldwide market value of carotenoids alone will be US$1,000 million (Del Campo et al., 2007). Some of the co-products fetch higher prices; for example, astaxanthin is about 3,000 times more expensive than the $1,000-per-ton crude oil (Cysewski and Lorenz, 2004). Although the payoffs for entrepreneurs are attractive, building biotech businesses based on a new, unproven technology poses more formidable challenges. Continuous production of vast quantities of algal biomass under optimal conditions is crucial in sustaining economically viable biofuel technology. Although fifty algal biofuel companies exist (http://aquaticbiofuel.com/2008/12/05/2008-the-year-of algae-investments/.), production on a commercial scale at competitive prices has not yet taken place (Pienkos and Darzins, 2009; St. John, 2009). One of the biggest challenges to commercial algal operations is to translate laboratory conditions to large scale, and most companies operate in "stealth" mode (Natural Resources Defense Council, 2009). To make it cost effective, Wijffels (2007) suggested that production costs must be reduced up to two orders of magnitude. When operating an algal biofuel production facility, plans should be in place to tackle unforeseen exigencies such as weather changes, and crashing of algal populations that could disrupt production and cause huge losses. As microalgae are renewable, sustainable, and affordable, their potential to produce biofuels is great if the current practices are cost competitive with petroleum diesel. Improvements in harvesting practices, extraction of biofuels, and conversion to co-products could bring down the production costs. Here we discuss the need to optimize various elements such as algal strains, cultivation, production costs, lipid variations, harvesting biomass, and genetic modification of microalgae to make microalgal biotechnology economically viable.

13.2 CULTIVATION

Approximately fifty species are utilized in biotechnology, mostly as biofeed. In these "traditional" species, manipulations of culture conditions (i.e., temperature, light, and nutrients) dramatically influence the yield of biomass. Long-term maintenance of algae may result in loss of algal vigor, resulting in "culture crashes" (Russell, 1974).

Cryopreservation of algae also contributes to death of cultures (Day and Harding, 2008). Such effects become obvious over long periods but are not evident in short periods—for example, the loss of B_{12} requirement in axenic cultures with long-term maintenance (Andersen, 2005). Continuous vegetative reproduction may lead to degeneration of cells, and such lost vigor can be restored by periodic sexual reproduction or the addition of organic base (Andersen, 2005). Growing multiple species may provide insight into their competition for nutrients and into the reproductive capacity of their vegetative stages (Riegman et al., 1996).

Cultivation of microalgae under laboratory conditions in defined sterile media, controlled temperature, and light influences their cost. For autotrophic microalgae, more than thirty kinds of media are used (Andersen, 2005; Subba Rao, 2009) and some of the commercially available nutrient stocks such as f/2 medium cost \$25 per liter. However, for outdoor mass cultivation systems and commercial developments, less-expensive media based on the enrichment of wastewater should be preferred. It would be necessary to carry out pilot experiments to critically evaluate the suitability of these media because of variations in their chemical composition. The use of wastewater, eutrophified water (Woertz et al., 2009; Kong et al., 2010, Park et al., 2011), secondary sewage (Orpez et al., 2009), dairy manure (Wang et al., 2010), swine manure effluent (Kebede-Westhead et al., 2006), farm effluents (Craggs et al., 2004), and commercial fertilizers such as Clewat-32™ (Ronquillo et al., 1997), Nualgi, SB07321(LM)M, Dyna-Gro™, and Miracle-Gro® may substantially reduce these costs while promoting vigorous algal growth.

Several designs for large-scale, flat-bed plane photobioreactors (PBRs) are available. Algae are also grown in a closed-loop, vertical or horizontal system of polyethylene sleeves, known as high density vertical growth (HDVG) systems, in greenhouses (Ugwu et al., 2008). Where land is at a premium, as at Schiphol Airport, Holland, it is proposed to construct an "ecobarrier"—a long tent parallel to the runway (Natural Resources Defense Council, 2009). Although a futuristic speculation, it is hoped that the ecobarrier supports algal cultivation and biofermentation technologies, and integrates transportation and landscape (Natural Resources Defense Council, 2009). However, an evaluation of the impact of temperature and light on their performance to sustain biomass levels is difficult because of the natural conditions. Because flat-plate photobioreactors suffer from a lack of uniform availability of light energy, it is suggested that circular-geometry bioreactors are better suited. Grobbelaar (2009) recommended closed PBRs because of their higher light utilization efficiencies, nutrient uptake, and biomass yield, and lower compensation light:dark ratios or respiratory losses, less contamination, and less water loss. A mean maximum of 98 g $m^{-2}d^{-1}$ with a maximum productivity of 170 g $m^{-2}d^{-1}$ is claimed by the Green Fuel reactor (Pulz, 2007), which can be attributed to a high surface volume ratio (SVR). For industrial purposes, algae are mass cultivated by the Israel-based Seambiotic in open-pond raceways ranging from 200 L to 1.2×10^6 L covering a 3,400-m^2 area.

Cultivation of algae under natural light and temperature is more cost-effective than under controlled laboratory conditions. *Chlamydomonas* sp., *Chlorella sorokiniana, Dunaliella tertiolecta, D. salina, Haematococcus pluvialis, Nannochloropsis* sp., *Phaeodactylum tricornutum, Porphyridium purpureum,*

P. cruentum, Scenedesmus obliquus, and *Synechocystis aquatilis* have been grown autotrophically in PBRs. Calculations with *Dunaliella* cultures showed that the use of a large, dense inoculum accelerates cell division with early attainment of stationary phase (Subba Rao, 2009). Such a shift saves time, which is desirable in a production process. Without negatively impacting growth rates, it is possible to attain a twofold increase in biomass in *Neochloris oleoabundans* by sequential increases in irradiance levels (Wahal and Vjamajala, 2010). Based on the geometry, fluid flow, and illumination on the biomass growth, Wu and Merchuk (2004) developed a triangular airlift reactor in which removal of CO_2 by two green algae (*Dunaliella parva* and *D. tertiolecta*) in a pilot-scale unit supplied with flue gases from a small power plant was 82.3 ± 12.5% on sunny days and 50.1 ± 6.5% on cloudy days.

The University of California, San Diego, designed a multi-stage algal bioreactor at the Scripps Institution of Oceanography (http://techtransfer.universityofcalifornia.edu/NCD/21141.html). This reactor provides light-limited growth, different or combined nutrient-controlled regimes, and can pre-amplify algal production to continuously inoculate existing pond or bioreactor systems.

Ben-Amoz (2009) reported that aeration of mass cultures with CO_2 flue gases enhances *Dunaliella* production from 2 to 20 g C $m^{-2}d^{-1}$ and could serve as an ideal and inexpensive nutrient source in commercial settings. Recent developments have substantially reduced production costs of microalgal dry biomass from $1,000 kg^{-1} in 1953 to a fraction of this ($0.17 to $0.29 kg^{-1}) when grown in wastewater (De Pauw et al., 1984). Israel-based Seambiotic produces *Dunaliella* for $17 kg^{-1} dry weight (DW) and by enrichment with CO_2 flue gases aims to produce *Nannochloropsis* for $1.00 kg^{-1} DW.

To lessen the costs associated with harvesting, Johnson (2009) provided "proof-of-concept" and cultured *Chlorella* sp. in the laboratory on attached solid polystyrene substrate on the bottom of a growth chamber. In dairy farm wastewater, *Chlorella* biomass reached production rates up to 3.2 g $m^{-2}d^{-1}$, comparable to suspended liquid cultures, and was harvested easily by scraping the solid surfaces. In addition to remediation of dairy manure water, an added advantage of this method was that the attached algal colonies served as inocula and eliminated the extra inoculation step (Johnson, 2009).

Although heterotrophic cultivation of algae could be cost-effective, only a few studies have been carried out. Growth of *Chlorella vulgaris* with the addition of the bacterium *Azospirillum brasiliense* in a heterotrophic regime, using glucose, yielded growth superior to that in cultures grown in autotrophic and mixotrophic regimes (Perez-Garcia et al., 2010). *Chlorella protothecoides* has been grown heterotrophically using an organic carbon source in an enclosed environment of fermenters ranging from 5 to 11,000 L capacity (Li et al., 2007). The PBRs and fermenters have the advantages of reduced contamination and evaporation, but are more expensive than open ponds and raceways. They can be utilized in the production of large volumes of inocula while switching over from an autotrophic mode to heterotrophic mode of cultivation.

Under the temperate climatic conditions of British Columbia (Canada), calculated base costs per liter of algal oil from raceways, closed PBRs, and fermenters correspond to \$2.66, \$7.32, and \$1.54, respectively (Alabi, 2009), and fermenters are recommended. In temperate regions, the climate limits cultivation of algae in open raceways to the warmer seasons. Alabi (2009) summarized production costs for mass cultivation of autotrophic algae as \$0.1 kg^{-1} to \$32 kg^{-1} compared to the heterotrophic cultivation (\$2.0 kg^{-1} to \$12 kg^{-1}). Algal-derived biofuel technology developed by Solix Biofuels costs about \$33 per gallon, or about US\$8 per liter (Kanellos, 2009). If biofuels are produced "dirt cheap" (Haag, 2007), production of fuel at an estimated cost of \$50 or less per barrel (Huntley and Redalje, 2007) would be economically viable.

13.3   NATIVE STRAINS, CONSORTIA OF SPECIES, AND EXTREMOPHILES

It is easy to find algae, but finding algae suitable for biotechnology is difficult. Currently, insufficient attention is paid to the selection of algal strains that could be cultivated inexpensively by growing them in wastewater and under ambient conditions of light and temperature. It is necessary for entrepreneurs of microalgal biotechnology to invest in selecting algal strains and optimizing their cultivation. The choice of commercial algal strains is of paramount importance and merits rigorous investigation. Local species are well adapted to local environmental conditions, and their utility contributes to more successful cultivation than nonnative species; for example, a consortium of *Actinastrum, Chlorella, Chlorococcum, Closterium, Euglena, Golenkinia, Micractinium, Nitzschia, Scenedesmus,* and *Spirogyra,* and two unidentified species concentrated from local ponds grew well at a dairy farm in municipal wastewater and yielded 2.8 g m^{-2} lipid day^{-1}, which would be equivalent to 11,000 L ha^{-1}y^{-1} (Pitman et al., 2011). Microalgal cultivation in wastewaters is cost effective in producing algal biomass for biofuel, and it also helps in the removal of nutrients (Craggs et al., 2011).

To date, few native species have been studied for their growth and photosynthetic efficiencies; with extremophiles, this is seldom the case. For example, photosynthetic rates of the extremophiles *Chlamydomonas plethora* and *Nitzschia frustule,* isolated from a semi-arid climate, approached their theoretical maxima corresponding to 22.8 and 18.1 mg C mg chl a^{-1} h^{-1} and high photosynthetic efficiencies (Subba Rao et al., 2005). Based on their specific growth rates at 10°C, 15°C, 25°C, and 30°C and threshold (I_0) and saturation (S) values of irradiance and saturation irradiance for growth, Kaeriyama et al. (2011) demonstrated the existence of physiological races in *Skeletonema* species isolated from Dokai Bay, Japan. Cultures of microalgae from tropical, subtropical, and semi-arid climates that may have unique physiological characteristics should be studied in detail. Of note, a marine diatom, *Navicula* sp. strain JPCC DA0580, and a marine green alga, *Chlorella* sp. strain NKG400014, isolated in Japanese ocean waters (Matsumoto et al., 2009) had a cell composition that yielded energy of 15.9 ± 0.2 MJ kg^{-1} and 26.9 ± 0.6 MJ kg^{-1}, respectively, which is equivalent to coal energy. Also of interest is the Strain B32 *Dunaliella* isolated from the Bay of Bengal, which yielded a maximum 0.68 pg carotene cell^{-1} while strain I3 yielded 17.54 pg carotene cell^{-1} (Keerthi et al., in press).

Extremophile algae stressed by high temperatures, light, salinity, and nutrients seem to have physiologically adapted to their harsh environmental conditions even under high irradiation, as evidenced by a chlorophycean microalga in the storage pools of nuclear reactors (Rivasseau et al., 2010). Because of their resilience, culturing these algae under ambient environmental conditions reduces the dependency on seasons for cultivation and the need to shut off operations during extreme climatic conditions. This will be cost-effective and enhance their utility in biotechnology. The thermo-acidophilic red alga *Galderia sulphuraria* isolated from environments with pH 0 to 4 pH and temperatures up to 56°C can survive both autotrophically and heterotrophically (Weber et al., 2004). This alga has a repertoire of metabolic enzymes with high potential for biotechnology. Its tolerance for high concentrations of cadmium, mercury, aluminum, and nickel supports its potential for bioremediation. The desert crusts seem to support extremophile members of five green algal classes; these unicellular algae growing under selective pressures of the desert appear to have high desiccation and photophysiology tolerance (Cardon et al., 2008). The extremophile cyanobacteria, mostly *Microcoleus* sp. living in the desert crust, are remarkably resistant to photo-inhibition, in contrast to *Synechocystis* sp. strain PCC 6803, and, within minutes of rehydration, recover their photosynthetic activity (Harel et al., 2004). Comparison of the extremophile *Chlamydomonas raudensis* Ettl UWO 241 isolated from an ice-covered Antarctic lake with its mesophilic counterpart *C. raudensis* Ettl. SAG 49.72 (SAG) isolated from a meadow pool in the Czech Republic, showed different abilities for acclimation (Pocock et al., 2011). The UWO 241 strain, unlike the other, relied on a redox sensing and signaling system for growth that bestows better success under stressful environmental conditions.

Nannochloris sp., isolated from the Great Salt Plains National Wildlife Refuge, grew in salinities from 0 to 150 PSU (practical salinity unit) and temperatures up to 45°C; growth and photosynthesis saturation were at 500 mol photons $m^{-2}s^{-1}$. Although the division rates in this alga were equal, in cells acclimated to low or high salinity and temperature, the former had a higher photosynthetic performance (P_{max}) than the latter (Major and Henley, 2008).

The extremophile *Coccomyxa acidophila* (pH < 2.5) accumulated more lutein (3.55 mg g^{-1}) when grown in urea (Casal et al., 2011). In another extremophile, *Chlamydomonas acidophila* (pH 2–3.5), stringent limitation of phosphate resulted in higher total fatty acid levels and lower percentages of polyunsaturated fatty acids (Spijkeman and Wacker, 2011). *C. acidophila* cultures grown on urea as a carbon source yielded high biomass levels (~20 g dry biomass $m^{-2}d^{-1}$) compared to ~14 g dry biomass $m^{-2}d^{-1}$ grown mixotrophically utilizing glucose as a carbon source (Cauresma et al., 2011). Mixotrophic growth of *C. acidophila* on glucose resulted in better accumulation of carotene and lutein (10 g kg^{-1} DW), the highest recorded for a microalga (Cauresma et al., 2011). In *Dunaliella salina* living under high light and salt stress, carotenogenesis shifted to higher salinity and increased substantially under nutrient-limiting conditions (Coesel et al., 2008); nutrient availability seems to control carotenogenesis and messenger-RNA levels. The extremophile (photopsychrophile) *Chlorella* sp. Strain BI isolated from Antarctica is unique in retaining the ability for dynamic short-term adjustment of light energy distribution between Photosystem II and Photosystem I, and can grow as a heterotroph in the dark (Morgan-Kiss et al., 2008).

13.4 VARIATIONS IN ALGAL PRODUCTION: CRUCIAL BUT IGNORED

Wide variations exist in units of measurement, and standardization is required with regard to the growth conditions of algae to permit comparison of outputs (Coronet, 2010). On a volume basis, biomass in several species of autotrophic algae varied considerably between 0.002 and 4 g $L^{-1}d^{-1}$ and 1.7 to 7.4 g $L^{-1}d^{-1}$ in the heterotrophic algae (Table 13.1); on an areal basis, values ranged from 0.57 to 150 g $m^{-2}d^{-1}$ (Table 13.1). The highest production of algal biomass (120 to 150 g $m^{-2}d^{-1}$) has been reported in PBRs under artificial light (Tsoglin and Gabel, 2000).

The success of microalgal biotechnology entrepreneurship depends on the optimization of biomass and production yields. It is necessary to establish to what extent these variations are intra-specific or inter-specific, whether or not these yields are based on optimal growth conditions, and how to prime the algal production. Between several species of *Dunaliella,* cell division rates ranged from 0.12 to 3.0 div d^{-1} (Subba Rao, 2009). Within the one species, *Chlorella sorokiniana,* biomass production rates (div d^{-1}) varied between 0.32 and 4.0 div d^{-1}; and in *Dunaliella teriolecta,* rates varied between 0.15 and 3.0 div d^{-1} (Subba Rao, 2009). Such variations could be due to differences in strains of isolates and/or culture conditions. Even in the most commonly used strain, *Neochloris oleoabundans* UTCC 1185, biomass varied between 0.03 and 1.50 g $L^{-1}d^{-1}$ (Table 13.2).

In *Dunaliella tertiolecta*, a green alga often used in biotechnology, Duarte and Subba Rao (2009) discussed the relationship between biomass (B determined as Chl-*a*), photosynthesis (P), and light energy I (μmol $m^{-2}s^{-1}$):

$$P^B = \{P^B_s[1 - \exp(-\alpha^B I/P^B_s)]\exp(-\beta^B I/P^B_s)\} + P^B_d$$

where P^B_s is the maximum potential photosynthesis in the absence of photoinhibition, and P^B_d is the intercept of the P–I curve on the *y*-axis and has the same units as P^B_m. In *D. teriolecta*, P^B_m varied between 3.3 and 7.43 mg C mg Chl-*a* h^{-1} (Duarte and Subba Rao, 2009). They showed that the photosynthesis and respiration activities were dependent on the light energy and the cell density; that is, over a 21-day period, gross production and respiration decreased by sevenfold and fourfold, respectively, at 42 μmol $m^{-2}s^{-1}$. The optimal light energy for photosynthesis ranged between 627 and 1,356 μmol $m^{-2}s^{-1}$. Also, the gross primary production:respiration ratio decreased with higher cell densities. It will be crucial in biotechnology operations to optimize the relationships among high biomass yields, photosynthetic efficiencies, and yield of bioactive compounds. These criteria are crucial and could greatly improve commercial algal harvest.

Grobbelaar (2010), while discussing the light energy relationships in algae, suggested that by optimizing light, photosynthetic yield could be doubled from 1.79 g (DW) $m^{-2}d^{-1}$ and pointed out that several factors determine volumetric yields of mass algal cultures. Furthermore, Grobbelaar pointed out that many biotechnology start-up companies make the mistake of simple extrapolation of controlled laboratory rates to large-scale outdoor production systems.

TABLE 13.1

Summary of Variations in Microalgal Biomass Production

Criteria	Example	Minimum	Maximum	Remarks	Ref.
Biomass	Haematococcus pluvialis	0.06	1.2	Bubble reactor 130 μmol photons m^{-2}s^{-1}	Garcia-Malea et al., 2006
	Haematococcus pluvialis	0.06	0.55	Tubular reactor	Garcia-Malea et al., 2005
	Haematococcus pluvialis		0.28		Esperanza Del Rio, 2005
	Haematococcus pluvialis		1.2	2,500 μmol photons m^{-2}s^{-1}	Hunt et al., 2010
	Chlorella sorokiniana		0.32	1,200 μmol photons m^{-2}s^{-1}	Janssen et al., 2003
	Chlorella sorokiniana		0.5		Chang and Yang, 2003
	Chlorella sorokiniana		1.8		Lee et al., 1996
	Chlorella sorokiniana		4		Gouveia et al., 2009
	Six microalgae	0.09	0.21		Chen et al., 2011
		0.04	0.37		Pitman et al., 2011
		0.03	0.48		Li et al., 2008
	Haematococcus oleoabundans		0.63	10 mM Sodium nitrate enrichment	Chen et al., 2011
	Haematococcus oleoabundans		0.4	5 mM Sodium nitrate enrichment	Chen et al., 2011
	Chlorella protothecoides	0.002–0.02	1.7–7.4	Phototrophic cultivation	Ben-Amoz, 2009
	Chlorella protothecoides		2	Heterotrophic cultivation	Ben-Amoz, 2009
Production	Dunaliella		120–150	Flue gas enriched	Tsoglin and Gabel, 2000
	Nannochloropsis sp.		20	Bioreactors, artificial light	Olaizola, 2000
	Haematococcus pluvialis		50–90		Chisti, 2007; Khan et al., 2009; Harun et al., 2010; Pitman et al., 2011; Chen et al., 2011
	Several species		0.91–38		
	20 species		0.57–130		Mata et al., 2010

Note: Variations in microalgal biomass (g L^{-1}d^{-1}) and production (g m^{-2}d^{-1}).

* All values are for autotrophic cultivation unless specified as heterotrophic.

TABLE 13.2

Variations in *Neochloris oleoabundans* UTCC 1185 Biomass

Species	Medium	Temperature (°C)	Light Intensity (μmol m^{-2}s^{-1})	Biomass Productivity (g L^{-1}d^{-1})	Ref.
Neochloris oleoabundans UTCC 1185	Bristol	26–30	150	0.03–0.15	Goueveia et al., 2009
Neochloris oleoabundans UTCC 1185	Erd Schreiber Soil extract	30	360	0.18–0.63	Li et al., 2009
Neochloris oleoabundans UTCC 1185	Bold modified	25	270	0.50–1.50	Pruvost et al., 2009
Neochloris oleoabundans UTCC 1185	Bristol modified	20	91–273	0.047–0.075	Wahal and Viamajlal, 2010

13.5   LIPID VARIATIONS: PHYSIOLOGICAL STATE

Microalgae produce tri- and di-glycerols, phospho- and glycol-lipids, and hydrocarbons (Chisti, 2007). Although claims regarding yield per acre are often exaggerated, third-generation microalgal biomass could yield 58,700 L biodiesel ha^{-1}y^{-1}, or even 90,000 L ha^{-1}y^{-1}, comparable to 53,200 L ha^{-1}y^{-1} (Weyer et al., 2010), an order of magnitude greater than the yields from first-generation biofuel crops (Chisti, 2007).

The mode of cultivation of algae is reflected in biomass and lipid yield (Table 13.3). Lipids as a percent of dry cell weight ranged between 1.9 and 75, and *Botryococcus braunii* yielded the highest percent (Malcata, 2010). Pienkos et al. (2011) summarized lipids (% of DW) in the range of 9.8% in cyanobacteria, 22.7% to 37.8% in diatoms, 25.5% to 45.7% in green microalgae, and 27.1% to 44.6% in other eukaryotic algae. Lipid production in autotrophic algae ranged from 0 to 2500 mg L^{-1}d^{-1}, and the highest was in *Chlorella protothecoides* (Chen et al., 2011). Areal production ranged from 0.57 to 38 g m^{-2}d^{-1}, and *Dunaliella salina* was the most productive (Mata et al., 2010).

Heterotrophy promotes faster growth and lipid accumulation. Compared to phototrophic cultures, cultures of *Chlorella protothecoides* grown heterotrophically had higher values of biomass productivity (1.7 to 7.4 g L^{-1}d^{-1}) and lipid productivity (732.7 to 3,701.1 mg L^{-1}d^{-1}), with lipid as percent dry cell weight ranging from 43% to 57.8% DW. *C. protothecoides*, when grown under heterotrophic conditions, yielded 55% lipid per cell dry weight (Xu et al., 2006). In mixotrophic cultures of *C. protothecoides* using glucose/acetate, higher levels of biomass (4.76 ± 1.50 g L^{-1}d^{-1}), biomass productivity (1.59 ± 0.50 g L^{-1}d^{-1}), and lipid productivity (0.25 g L^{-1}d^{-1}) were obtained; but because the cost of the raw materials was unacceptable, glycerol and acetate were used as carbon sources (Heredia-Arroyo et al., 2010). With glycerol, the corresponding values were 3.97, 0.93, and 0.19 g L^{-1}d^{-1} (Heredia-Arroyo et al., 2010). However, in phototrophic cultures of *C. protothecoides,* corresponding

TABLE 13.3
Summary of Inter-specific and Intra-specific Variations in Microalgal Lipid

Criteria	Example	Min.	Max.	Remarks[a]	Ref.
Lipid (% of cell dry weight)	*Chlorella protothecoides*	43	57.8	Heterotrophic cultivation	Chen et al., 2011
	32 species	11	23		Chen et al., 2011
	44 species	5	67.8		Chen et al., 2011
	Several species	5	63		Mata et al., 2010
		5	63		Chisti, 2007; Khan et al., 2009; Harun et al., 2010; Pitman et al., 2011; Chen et al., 2011
	Scenedesmus dimorphus	6			Gouveia and Oliveira, 2009
	Botryococcus braunii	25	75		Chisti, 2007
	Schizochrtrium sp.	77			Chisti, 2007
	Scenedesmus obliquus	11	55		Gouveia and Oliveira, 2009
	Chlorella vulgaris	14	55		Gouveia and Oliveira, 2009
	Chlorella protothecoides	23	55		Gouveia and Oliveira, 2009
	Neochloris oleoabundans	35	65		Gouveia and Oliveira, 2009
	Nannochloropsis oculata	31		Log phase	Chiu et al., 2009
	Nannochloropsis oculata	40		Early stationary phase	
	Nannochloropsis oculata	50		Stationary phase	
	Nannochloropsis sp.	22	60		Rodolphi et al., 2009
Lipid production (mg L⁻¹d⁻¹)	*Chlorella protothecoides*	733	3701.1	Heterotrophic cultivation	Chen et al., 2011
	Chlorella protothecoides	0	5.4		Chen et al., 2011
	32 species	0	178.8		Chen et al., 2011
	Chlorella protothecoides	0	1214		Huerlimann et al., 2010

[a] All values are for autotrophic cultivation unless specified otherwise.

(Continued)

TABLE 13.3 (Continued)

Summary of Inter-specific and Intra-specific Variations in Microalgal Lipid

Criteria	Example	Min.	Max.	Remarks[a]	Ref.
	27 species	0	133		Huertlimann et al., 2011
	Several species	3	2500		Chisti, 2007; Khan et al., 2009; Harun et al., 2010; Pitman et al., 2011; Chen et al., 2011
	Nannochloropsis sp.	30	86.3		Chiu et al., 2009
	44 species	10.3	142		Mata et al., 2010
pg lipid cell^{-1}	4 species	0	29.108		Huertlimann et al., 2010

[a] All values are for autotrophic cultivation unless specified otherwise.

values were 0.002 to 0.02 g $L^{-1}d^{-1}$ biomass, 0.2 to 5.4 mg $L^{-1}d^{-1}$ lipid, and 11% to 23% lipid dry cell weight (Chen et al., 2011). Twenty-one other phototrophic species had a range of biomass production rates from 0.02 to 0.53 g $L^{-1}d^{-1}$, 0.2 to 178.8 mg $L^{-1}d^{-1}$ lipid, and 5.1% to 67.8% lipid dry cell weight. Calculation of lipids on a cell basis also varied from 0.068 to 29.11 pg $cell^{-1}$ (Huerlimann et al., 2010).

The data on lipid variations given by Chisti (2007), Khan et al. (2009), Harun et al. (2010), Basova (2005), Huerlimann et al. (2010), Mata et al. (2010), Malcata (2010), and Chen (2011) summarized in Table 13.3 show that wide inter- and intra-specific variations in lipid levels as percent dry cell weight exist. For example, the lowest (6%) was in *Scenedesmus dimorphus* (Gouveia and Oliveira, 2009), compared to 75% in *Botryococcus braunii* (Chisti, 2007) and 77% in *Schizochrtrium* sp. (Chisti, 2007). Gouveia and Oliveira (2009) reported a wide range of values within the same species: 11% to 55% in *Scenedesmus obliquus*, 14% to 56% in *Chlorella vulgaris*, 23% to 55% in *Chlorella prototothecoides,* and 35% to 65% in *Neochloris oleoabundans.* Chisti (2007) reported 25% to 75% in *Botryococcus braunii.* These variations could be attributed to variations in the physiological state of the cells; marked differences in the lipid were noticed in cultures harvested in logarithmic, late logarithmic, and stationary phases of *Nannochloropsis* sp., *Isochrysis* sp., *Tetraselmis* sp., and *Rhodomonas* sp. (Huerlimann et al., 2010). Results of Chiu et al. (2009) corroborate that lipids vary with the phase of growth of the alga *Nannochloropsis oculata*; lipids were 30.8% in log phase cultures, 39.7% in early stationary phase, and 50.4% in stationary phase cells (Chiu et al., 2009). In *Nannochloropsis* sp., the lipid as percent dry cell weight ranged from 21.6 to 60 (Rodolfi et al., 2008; Chiu et al., 2009), and their production rates correspond to 30 mg $L^{-1}d^{-1}$ and 86.3 mg $L^{-1}d^{-1}$, respectively. Sturm and Lamer (2011), based on energy evaluation from wastewater algal biomass production, concluded that if the lipid in dry biomass from the field is less than 10%, compared to 50% to 60% in laboratory-scale reactors, it would be better to use the biomass as a combustible source of viable energy.

13.6 BIOCHEMICAL MANIPULATION: HIGHER YIELDS

Chemical manipulations of algae are reflected in their biochemical constituents. Nitrogen starvation increased lipid production from 117 to 204 mg $L^{-1}d^{-1}$ (Rodolfi et al., 2008). By manipulating the nutrients in *Scenedesmus obliquus*, up to 58.3% lipid was attained, which was five- to tenfold higher than controls (Mandal and Mallick, 2009). It is of interest to note that carotenoids increased only in *Dunaliella salina* as the salinity increased (Gómez et al., 2003; Coesel et al., 2008). The carotenoid levels (mg L^{-1}) corresponded to 6.9, 10.8, and 12.9 mg L^{-1} in Provosoli medium of 1 M, 2 M, and 3 M sodium chloride (NaCl), respectively; in an artificial medium, they were more pronounced and were 8, 12.9, and 29.5 mg L^{-1} in 1 M, 2 M, and 3 M NaCl, respectively. Takagi et al. (2006) showed that the salt content of the medium could also be a stressor in *Dunaliella*. In the initial stages of cultures, when the NaCl was increased from 0.5 M (equivalent to seawater) to 1.0 M, lipid increased by 67%; when mid- or late-log phase cultures were subjected to a similar stress, cellular lipid increased to 70%. So while harvesting cells for biotechnological applications, the strain and physiological state of the algae play critical roles in determining output.

In addition to production rates, variations in biochemical profiles must be considered for optimization of harvesting. Carbohydrates, proteins, lipids, and fats are known to vary with the medium used among seven species of marine microalgae (Fernández-Reiriz et al., 1989) and in sixteen species of microalgae commonly used in aquaculture (Brown, 1991). The medium used (Walne, ES, f/2, and Algal-1) for cultivation also influenced the biochemical profiles of four species (Fernández-Reiriz et al., 1989).

Through biochemical manipulation, lipid synthesis can be regulated; this involves imposing a physiological stress such as nutrient starvation to channel metabolic processes toward lipid accumulation. In experiments by Li et al. (2008), cultures of *Neochloris oleoabundans* were supplied with sodium nitrate, urea, and ammonium bicarbonate as the nitrogen source; only at lower levels of sodium nitrate did cellular lipid increase. Co-limitation for inorganic phosphorus and carbon dioxide in *Chlamydomonas acidophila* Negoro resulted in high photosynthetic rates and also in a mismatch between photosynthesis and growth rates in phosphorus-limited cultures (Spijkerman, 2010). In *Monodus subterraneus* when phosphate was decreased from 175 to 52.5, 17.5, or 0 μM, cellular lipid increased (Khozin-Goldberg, and Cohen, 2006). Limitation of nitrogen in cultures of the green alga *Scenedesmus obliquus* resulted in an increase of lipid from 12.7% to 43% of cell dry weight (DW) (Mandal and Mallick, 2009); a deficiency of phosphate increased lipid to 29.5% (DW). Lipids in nitrogen-limited *Chaetoceros mulleri* increased five- to sevenfold compared to nitrogen-replete cultures (McGinnis et al., 1997). Results obtained with *Nannochloropsis oculata* and *Chlorella vulgaris* (Converti et al., 2009) confirmed such an impact of nitrogen limitation. Lipid production is enhanced to 90 kg ha^{-1}d^{-1} by a two-stage culture system that involves raising high-density cultures under optimal conditions initially and then transferring them to a nitrogen-deficient medium (Rodolfi et al., 2008).

In *Dunaliella salina* cultures, an increase in CO_2 from 2% to 10% increased lipid production by 170% in 7 days (Muradyan et al., 2004). In *Chlamydomonas vulgaris,* the addition of 1.2×10^{-5} M Fe^{3+} not only suppressed cell growth initially, but also enhanced the accumulation of lipids up to 56.6% DW. Furthermore, the accumulation of lipids occurred earlier during the stationary phase (Liu et al., 2008). To enhance the yield of microalgal biomass, rigorous experiments should be carried out to establish the impact of several micronutrients, such as selenium and boron. Optimized growth of commercial algae should account for the effects of manipulations in nutrients, temperature, and chemical composition of media.

13.7   HARVESTING

Because of their geometry, secretion of mucilage and variations in cell weight, microalgae must be harvested in a species-specific, nongeneric manner (Benemann, 2008). Considerable process-oriented research is needed to optimize methods of algal harvest, lipid extraction, and purification of by-products. Flocculation is widely used for algal harvest using various salts (Grima et al., 2003) such as aluminum, iron, potassium, zinc, chitosan, extracellular polymeric substances, bioflocculants such as *Paenibacillus* + aluminum sulfate, and organic cationic polymers. Co-bioflocculation (*Nannochloris* + diatoms) is also used, but necessitates extra effort to grow another alga. Pressure filtration (10 PSIG [pounds per square inch gage]) through four conical felt media bags (1 μm), ultrasonication and grinding, cross-flow microfiltration/ultrafiltration, and continuous foam separation are some of the other methods used for algal harvest. These techniques are labor intensive, expensive, and inefficient, with yields in the range of 30% biomass. Coagulation in the presence of chemicals is an alternative method for harvesting algae. *Scenedesmus subspicatus, Selenastrum capricornutum,* and *Nannochloropsis* sp. exposed to 5 mg L^{-1} barium concentrations bio-accumulate up to 88% to 99% of barium within 10 days (Theegala et al., 2001). Further treatment with 200 mg L^{-1} ferric chloride facilitated harvest of the metal-laden microalgae with an efficiency of nearly 99%.

Tetraselmis suecica can be concentrated up to 148 times using tangential flow filtration (TFF) and up to 357 times with polymer flocculation (PF); TFF requires a high initial capital investment and consumes 2.06 kWh m^{-3} while PF requires low initial investment with energy consumption in the range of 14.81 kWh m^{-3} (Danquah et al., 2009). The payback period, an important criterion for the investor, is 1.5 years for TFF and 3 years for PF. Passive and active immobilization techniques hold promise for harvesting high-value molecules such as storage products, antibiotics, hydrocarbons, hydrogen, and glycerol, and should be explored (Lebeau and Robert, 2006). Algaeventure Systems (AVS) has been developing an AVS Harvester that uses conveyor belts of capillaries to concentrate and dry *Chlorella* cultures. The estimated processing cost is $1.92 per ton compared to $875 per ton by centrifugation. One of the drawbacks is the required algal concentration of 3 g L^{-1}; improvements are being pursued to increase harvesting efficiency.

An aqueous and a biocompatible organic phase (dodecane) bioreactor exist to extract β-carotene from *Dunaliella salina* cells. The organic phase continuously removes β-carotene ("milked") from the cells with greater than 55% efficiency, and productivity is 2.45 mg $m^{-2}d^{-1}$, which is much higher than that of commercial plants (Hejazi et al., 2004). Several other methods of harvesting microalgae from liquid cultures are being developed, looking for a breakthrough to drastically reduce harvesting and dewatering costs. Although details have not been published, mention should be made of the following:

1. Pretreatment of algae that involves application of 10 to 30 kV cm^{-1} electrical pulses for 2 to 20 μs to an algal slurry to rupture the cell walls and to release biodiesel compounds such as methyl hexadecanoate (Diversified Technologies at the University of Galway, Ireland).
2. Usage of amphiphilic solvents, such as acetone, methanol, ethanol, isopropanol, butanone, dimethyl ether, or propionaldehyde, to separate out the proteins and carbohydrates from the lipids (Aurora Algae).
3. Harvesting, dewatering, and drying system utilizing surface physics and low-energy capillary action (Algaeventure Systems).
4. Single-step and live extraction of lipids (Origin Oil, James Cook University).
5. Hydrothermal liquefaction or thermal depolymerization (New Oil Resources).

6. A two-step catalyst-free algal biodiesel production process, using wet algal biomass and bypassing the drying and solvent extraction steps (University of Michigan).
7. Acoustic-focusing technology that generates ultrasonic fields that concentrate algal cells into a dense sludge and extract oil (Solix).

Additionally, Kleinegris et al. (2011) have discussed product excretion, cell permeabilization, and cell death as mechanisms to extract microalgal products. They propose using two-phase systems that could circumvent the step of harvesting algal cells while the product is extracted in situ and prepared for downstream processing.

13.8   GENETIC MODIFICATION OF ALGAE

Genome analysis is available for only four unicellular algae: *Chlamydomonas reinhardtii, Cyanidoschyzon merolae, Ostreococcus tauri,* and *Thalassiosira pseudonana* (Misumi et al., 2008). Genetic modification (GM) of microalgae holds promise as a strategy to attain higher lipid yields while concurrently generating value-added products (Jin et al., 2003; León and Fernández, 2007; Gressel, 2008). Although several hundred strains of microalgae have been cultured, detailed investigation of cellular physiology and biochemistry is limited to fewer than thirty species. Fewer still are the algal strains that have been studied at the genomic level. Genetic transformation of microalgae has been constrained by the presence of rigid cell walls (Rosenberg et al., 2008). However, using a plethora of techniques such as bombardment, electroporation, and treatment with silicon whiskers and glass beads, several species have been modified genetically (León and Fernández, 2007), including *Amphidinium* sp., *Anabaena* sp., *Chlamydomonas* sp., *Chlorella ellipsoidea, C. kessleri, C. reinhardtii, C. sacchrophila, C. sorokiniana, C. vulgaris, Cyclotella cryptica, Cylindrotheca fusiformis, Dunaliella salina, Euglena gracilis, Haematococcus pluvialis, Navicula saprophila, Phaeodactylum tricornutum, Porphyridium* sp., *Symbiodinium microadriaticum, Synechocystis* sp., *Thalassiosira weisflogii,* and *Volvox carteri.* The red alga *Cyanidoschyzon merolae* and the euglenoid *Euglena gracilis* have also been genetically transformed (Rosenberg et al., 2008). We agree with Pienkos et al. (2011), who suggest that through genetic engineering a few "designer algal strains" that have all the properties needed for large-scale biotechnology should be developed, and more research must be carried out in parallel with natural strains to fully understand their physiological functioning. Such modifications can impart properties to improve yield. For instance, Li and Tsai (2008) demonstrated that the microalga *Nannochloropsis oculata,* which was codon-optimized to produce bovine lactoferricin (LFB) fused with a red fluorescent protein (DsRed), has bactericidal defense against *V. parahaemolyticus* infection in the shrimp digestive tract.

The utility of engineered microalgae for augmented lipid biosynthesis, conversion from autotrophy to heterotrophy, enhancing photosynthetic conversion efficiency and expression of recombinant proteins is gaining prominence (Rosenberg et al., 2008). While it is possible to enhance lipid synthesis through cloning acetyl-CoA carboxylase (ACC) genes in yeast, fungi, bacteria, and a few higher plants, there was no change in lipid content of a similarly engineered diatom *Cyclotella cryptica* (Dunahay et al., 1995; Dunahay et al., 1996). Three possible strategies exist for enhanced lipid production: biochemical engineering (BE), genetic engineering (GE), and transcription factor engineering (TFE). BE approaches are currently the most widely established in microalgal lipid production (Courchesne et al., 2009).

Radakovits et al. (2010) discussed the potential of manipulating the central carbon metabolism in eukaryotic microalgae through genetic engineering to enhance lipid production. They suggested that it should be possible to increase production of not only carbon storage compounds, such as TAGs and starch, but also designer hydrocarbons that may be used directly as fuels.

Another possibility is to engineer the light-harvesting antennae in autotrophic algae. Smaller antennae lead to greater photosynthetic efficiency (Mitra and Melis, 2008); mutating genes that control antennae biogenesis is a possible mechanism for enhancing photosynthetic efficiency (Scott et al., 2010). Possibilities exist to improve solar energy conversion efficiency from the present the 1–4% to 8–12% to realize fully the potential of microalgal co-production systems in *Chlamydomonas perigranulata, C. reinhardtii, Chlorella vulgaris, Cyclotella* sp., *Dunaliella salina, Scenedesmus obliquus,* and *Synechocystis* PCC 6714 (Stephens et al., 2010).

A mechanistic model developed by Flynn et al. (2010) explores cellular chlorophyll and photosynthetic efficiency to optimize commercial algal biomass production. The model predicts that genetically modified strains with a large antenna size, indicated by a low Chl:C ratio, are more suitable for commercial biofuel production than strains selected from nature. However, for the generation of hydrogen and hydrocarbons as biofuels, smaller light-collecting antennae seem to be more efficient in *Botryococcus braunii* (Eroglu and Melis, 2010). Three races (Race A, B, and L) of the strain *Botryococcus braunii* are recognized (Banerjee et al., 2002); these races are regarded as a potential source of renewable fuel with yields of hydrocarbons reaching up to 75% of algal dry mass. A *Botyrococcus* Squalene Synthase (BSS) gene from a Race B variant of *B. braunii* has been sequenced, amplified as a 1,403-bp fragment, and expressed as a heterologous protein in *E. coli* BL21 cells. Following Isopropyl-β-D-thiogalactoside (IPTG) induction, recombinant squalene synthase activity was detected, suggesting that a key hydrocarbon synthesis gene from a commercial alga can be isolated and cloned into a heterologous expression system. This opens the door for large-scale hydrocarbon synthesis in more amenable systems such as *E. coli* and may help reduce the problems associated with the viscous nature of *Botyrococcus* cultures (Banerjee et al., 2002).

13.9 SUMMARY

Innovative ways to optimize maximum microalgal biomass production and technological advances for transesterification would be necessary to make microalgae more cost effective for biodiesel production and to sustain an economically viable microalgal biotechnological industry (Figure 13.1). Improvements at various intermediary stages of culturing, selection of strains of algae, harvesting, and extraction of bio-fuel production and co-products could bring down the production costs. Norsker et al. (2011) state that by optimizing irradiation conditions, mixing, photosynthetic

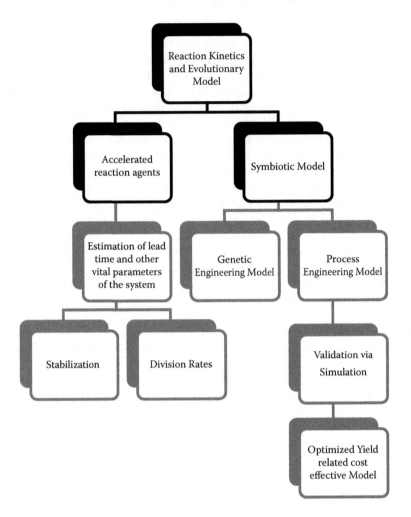

FIGURE 13.1 Schematic model for microalgal biotechnology.

efficiency, growth media, and CO_2 costs, the overall cost of production could be reduced to Euro 0.68 per kilogram, which would be economically acceptable for using algae as feedstock for biodiesel and chemicals. Alternative metabolic pathways such as heterotrophy and mixotrophy should be explored to maximize algal growth without a shift to energetically inefficient metabolism. Service (2008) states that algae grown in dark stainless steel fermenters convert sugars to oils more efficiently. Heterotrophic and mixotrophic cultivation of microalgae in fermentation systems for commercial viability should be explored (Gladue and Maxey, 1994; Xu et al., 2006; Chi et al., 2007).

Microalgae hold great potential as a source of cheap and environmentally friendly biofuel. The total annual production of microalgae in 2004 was 5,000 tons, with global sales worth about US$1.25 billion (Pulz and Gross, 2004). However, we believe that comprehensive evaluation of select species from an integrated perspective would be of greatest benefit to commercial operations. Although Serrano's quote (Serrano, 2010) that, "We are still like the Wright Brothers, putting pieces of wood and paper together" is in a different context, it is apt here. The rigor of microalgal biofuel research, coupled with its interdisciplinary nature, suggests that a comprehensive modeling strategy, one that accounts for numerous culture and harvest parameters and optimizes industrial processes from a perspective of cost, would be of great value. Simulation models that incorporate elements of nutrient systems, ideal culture conditions, and harvest of multiple products such as fuels and high-value nutraceuticals and/or recombinant proteins would be instrumental in the development of a viable bio-economy. Brown (2009) pointed out that as mass cultivation of algae for biofuels per se may not sustain microalgal technology, attention should be paid to non-fuel products and co-products as well. These co-products include carotenoids, phycolbiliproteins, astaxanthin, and eicosapentaenoic acid; additionally, algal biomass waste could be used as fertilizer (Donovan and Stowe, 2009).

Various processes are involved in this modeling activity. As Malcata (2010) observed, modeling exercises, instead of empirical approaches, should have biological meaning for which specific experimental data should be obtained on the optimum versus enhanced growth, metabolic cycles, assimilation efficiencies, that is, conversion of substrate into reserves, accumulation, and product sysnthesis/excretion. Scott et al. (2010) commented that there is an inadequacy of established background knowledge in this area, and there is a need to integrate biology and engineering.

The central theme rests on the predictive aspects of modeling that enable one to determine the exact quantities of the envisaged end product together with co-products. To estimate the actual quantities, we require appropriate input data regarding culture conditions, harvest efficiencies, and yield of co-products, as outlined above. The effective price for the microalgae-derived biofuels can be calculated by optimizing the cost functional involving several variables under appropriately formulated constraints. Results obtained from all stages of the process constitute the vital parameters in the mathematical model. As the process is dynamic in character, time delays do occur in a natural way, and these delays account for the process lead-time. We need to estimate these time delays, maintaining the stability of the corresponding delay-free systems. Division rates of the reaction mechanisms play a vital role in the process of restoring and/or maintaining the stability of the processes. Simulations based on realistic data will grossly help in the validation of these models. Thoroughly validated models are utilized for predicting the optimal cost of biofuel under conditions where lipid yields are maximum.

Williams and Laurens (2010) argued that a fundamental change in the approaches to production is needed, and that "biofuel-only" options may not be economically viable. They showed that 30% to 50% of primary production is lost in the production of protein and lipid, and that if lipid production is increased, then production of other valuable co-products is reduced. These authors argue that the availability of nutrients such as phosphorus and nitrogen, delivery of CO_2, and the energy costs associated with sterilization and recycling of spent culture water and removal of biological contaminants, pathogens, and predators would escalate production of microalgal biomass and could be "show-stoppers."

In conclusion, microalgal biotechnology has made rapid advances in the mass cultivation of algae and their application toward biofeed, biopharmacy, biofuel, bio-remediation, bioactive compounds, and space research. However, fewer than fifty species are utilized, while thousands remain unexplored. The potential roles of microalgae in genetic engineering and nanotechnology have increased the prospects for the next generation of "designer microalgae." To establish algal biotechnology as an economically viable enterprise, concerted research is needed to (1) develop inexpensive media through enrichment of wastewater; (2) isolate and culture new strains of high-yielding microalgae, preferably a consortium of extremophiles; (3) improve production systems; (4) enhance biochemical and metabolic pathways through genetic engineering; and (5) improve harvesting techniques.

Additionally, attention should be given to high-value natural and recombinant products that can be extracted from algae to enhance the profitability of biofuel operations. Simulation models will serve as the foundation for industrial processes that optimize wastewater treatment systems, nutrient levels, and strategies for harvest and extraction of bioactive compounds. A robust bio-economy built on a platform of innovative microalgal technologies requires a cross-disciplinary approach among biologists, biotechnologists, molecular biologists, biochemists, engineers, chemists, bioreactor manufacturers, aquaculturists, and modelers.

ACKNOWLEDGMENTS

We are grateful to Professor Faizal Bux and Dr. Taurai Mutanda, Institute for Water and Wastewater Technology, Durban University of Technology, South Africa, for inviting us to contribute this chapter. We are most grateful to Professor John Beardall, Monash University, Clayton, Victoria, Australia, for constructive review of the manuscript. We thank Bala T. Durvasula and Dr. Ivy Hurwitz for their help with formatting the manuscript.

The research of V. Sree Hari Rao is supported by the Foundation for Scientific Research and Technological Innovation (FSRTI), a Constituent Division of the Sri Vadrevu Seshagiri Rao Memorial Charitable Trust, Hyderabad, India.

REFERENCES

Alabi, A.O. (2009). Microalgal Technologies and Processes for Biofuels/Bioenergy Production in British Columbia. The British Columbia Innovation Council, Vancouver, pp. 1–74.

Allen, E.J., and Nelson, E.W. (1910). On the artificial culture of marine plankton organisms. *Journal of the Marine Biological Association of the United Kingdom*, 8: 421–474.

Andersen, R.A. (2005). *Algal Culturing Techniques*. Academic Press, Amsterdam.

AquaticBiofuel.com (2008). 2008 the Year of Algae Investments (December 5, 2008), accessed June 30, 2009, http://aquaticbiofuel.com/2008/12/05/2008-the-year-of-algae-investments/.

Banerjee, A., Sharma, R., Chisti, Y., and Banerjee, U.C. (2002). *Botryococcus braunii*: A renewable source of hydrocarbons and other chemicals. *Critical Reviews in Biotechnology,* 22: 245–279.

Basova, M.K. (2005). Fatty acid composition of lipids in microalgae. *International Journal on Algae*, 7: 33–57.

Benemann, J.R. (2008). *NREL-AFOSR Workshop*, Algal Oil for Jet Fuel Production; Arlington, VA, 19 February 2008.

Benemann, J.R. (1993). Utilization of carbon dioxide from fossil fuel-burning plants with biological systems. *Energy Conversion and Management*, 34: 999–1004.

Ben-Amoz, A. (2009). Bioactive compounds: Glycerol production, carotenoid production, fatty acids production. In *The Alga Dunaliella Biodiversity, Physiology, Genomics and Biotechnology* (Eds. Ben-Amotz, A., Polle, J.E.W., and Subba Rao, D.V.). Science Publishers, Enfield, NH, pp. 189–207.

Brown, M.R. (1991). The amino acid and sugar composition of 16 species of microalgae used in mariculture. *Journal of Experimental Marine Biology and Ecology*, 145: 79–99.

Brown, P. (2009). Algal Biofuels Research, Development, and Commercialization Priorities: A Commercial Economics Perspective. Energy Overviews. [Online] ep Overviews Publishing, Inc, 22. 06. 2009. http://www.epoverviews.com/oca/Algae%20Biofuel%20Development%20Prioritie.

Cardon, Z.G., Gray, D.W., and Lewis, L.A. (2008). The green algal underground: Evolutionary secrets of desert cells. *Bioscience*, 58: 114–122.

Casal, C.C., Cauresma, M.M., Vega, M.J.M., and Vilches, C.C. (2011). Enhanced productivity of a lutein-enriched novel acidophile microalga grown on urea. *Marine Drugs*, 9: 29–42.

Cauresma, M., Casal, C., Forjanb, E., and Vilches, C.C. (2011). Productivity and selective accumulation of carotenoids of the novel extremophile microalga *Chlamydomonas acidophila* grown with different carbon sources in batch systems. *Journal of Industrial Microbiology and Biotechnology*, 38: 167–177.

Chang, E.H., and Yang, S.S. (2003). Some characteristics of microalgae isolated in Taiwan for biofixation of carbon dioxide. *Botanical Bulletin of Academia Sinica*, 44: 43–52.

Chen, C.Y., Yeh, K.L., Aisyah, R., Lee, D.J., and Chang, J.S. (2011). Cultivation, photobioreactor design and harvesting of microalgae for biodiesel production: A critical review. *Bioresource Technology*, 10: 71–81.

Chi, Z.Y., Pyle, D., Wen, Z.Y., Frear, C., and Chen, S.L. (2007). A laboratory study of producing docosahexaenoic acid from biodiesel-waste glycerol by microalgal fermentation. *Process Biochemistry*, 42: 1537–1545.

Chisti, Y. (2007). Biodiesel from microalgae. *Biotechnology Advances*, 25: 294–306.

Chiu, S., Kao, C.C., Tsai, M., Ong, S., Chen, C., and Lin, C. (2009). Lipid accumulation and CO_2 utilization of *Nannochloropsis oculata* in response to CO_2 aeration. *Bioresource Technology*, 100: 833–838.

Coesel, S.N., Baumgartner, A.C., Teles, U.M., Ramo, A.A., and Henriques, N.M. (2008). Nutrient limitation is the main regulatory factor for carotenoid accumulation and for Psy and Pds steady state transcript levels in *Dunaliella salina* (Chlorophyta) exposed to high light and salt stress. *Marine Biotechnology*, 10: 602–611.

Converti, A., Casazza, A., Ortiz, E., Perego, P., and Borghi, M. (2009). Effect of temperature and nitrogen concentration on the growth and lipid content of *Nannochloropsis oculata* and *Chlorella vulgaris* for biodiesel production. *Chemical Engineering and Processing*, 48: 1146–1151.

Coronet, J.F. (2010). Calculation of optimal design and ideal productivities of volumetrically lightened photo bioreactors using the constructal approach. *Chemical Engineering Science*, 65: 985–998.

Courchesne, N.M.D., Parisien, A., Wang, B., and Lan, C.Q. (2009). Enhancement of lipid production using biochemical, genetic and transcription factor engineering approaches. *Journal of Biotechnology*, 141: 31–41.

Craggs, R.J., Sukias, J.P., Tanner, C.T., and Davies-Colley, R.J. (2004). Advanced pond system for diary-farm effluent treatment. *New Zealand Journal of Agricultural Research*, 47: 449–460.

Craggs, R.J., Heubeck, S., Lundquist, T.J., and Benemann, J.R. (2011). Algal biofuels from wastewater treatment high rate algal ponds. *Water Science and Technology*, 63: 660–665.

Cysewski, G.R., and Lorenz, R.T. (2004). Industrial production of microalgal cell-mass and secondary products—Species of high potential: *Haematococcus*. In *Handbook of Microalgal Culture*. Wiley-Blackwell, United Kingdom, pp. 281–288.

Day, J.G., and Harding, K.K. (2008). Cryopreservation of algae. In *Plant Cryopreservation: A Practical Guide* Biomedical and Life Sciences. Plant Cryopreservation Section II (Ed. B.M. Reed), Springer, New York, pp. 95–116. doi: 10.1007/978-0-387-72276-4_6.

De Pauw, N., Morales, J., and Persoone, G. (1984). Mass culture of microalgae in aquaculture systems: Progress and constraint. *Hydrobiologia*, 116/117: 121–134.

Duarte, P., and Subba Rao, D.V. (2009). Photosynthesis – Energy relationships in *Dunaliella*. In *The Alga Dunaliella Biodiversity, Physiology, Genomics and Biotechnology* (Eds. Ben-Amotz A., Polle, J.E.W., and Subba Rao, D.V.), Science Publishers, Enfield, NH, pp. 209–229.

Dunahay, T.G., Jarvis, E.E., and Roessler, P.G. (1995). Genetic transformation of the diatoms *Cyclotella cryptica* and *Navicula saprophila*. *Journal of Phycology*, 31: 1004–1012.

Dunahay, T. G., Jarvis, E.E., Dai, S.S., and Roessler, P.G. (1996). Manipulation of microalgal lipid production using genetic engineering. *Applied Biochemistry and Biotechnology*, 57–58(1): 223–231.

Danquah, M.K., Ang, L., Uduman, N., Moheimani, N., and Forde, G.M. (2009). Dewatering of microalgal culture for biodiesel production: Exploring polymer flocculation and tangential flow filtration. *Journal of Chemical Technology and Biotechnology*, 84: 1078–1083.

Del Campo, J.A., Garcia-González, M., and Guerrero, M.G. (2007). Outdoor cultivation of microalgae for carotenoid production: Current state and perspectives. *Applied Microbiology and Biotechnology*, 74: 1163–1174.

Donovan, J., and Stowe, N. (2009). Is the Future of Biofuels in Algae? [Online] Renewable Energy World, 12 06 2009. http://www.renewableenergyworld.com/rea/news/article/2009/06/is-the-futu.

Eroglu, E., and Melis, A. (2010). Extracellular terpenoid hydrocarbon extraction and quantitation from the green microalgae *Botryococcus braunii* var. *Showa*. *Bioresource Technology*, 101: 2359–2366.

Esperanza, D.R.F., Gabriel, A.M., Carmen, G.M., Joaquín, R., Emilio, M.G., and Miguel, G.G. (2005). Efficient one-step production of astaxanthin by the microalga *Haematococcus pluvialis* in continuous culture. *Biotechnology and Bioengineering*, 91: 808–815.

Fairley, P. (2011). Introduction: Next generation biofuels. *Nature*, 474: S2–S5. doi:10.1038/474S02a

Fernandez-Reiriz, M.J., Perez-Camacho, A., Ferreiro, M.J., Blanco, J., Planas, M., Campos, J., and Labarta, U. (1989). Biomass production and variation in the biochemical profile (total protein, carbohydrates, RNA, lipids and fatty acids) of seven species of marine microalgae. *Aquaculture*, 83: 17–37.

Flynn, K.J., Greenwell, H.C., Lovitt, R.W., and Shields, R.J. (2010). Selection for fitness at the individual or population levels: Modelling effects of genetic modifications in microalgae on productivity and environmental safety. *Journal of Theoretical Biology*, 263: 269–280.

García-Malea, M.C., Brindley, C., Del Rio, E., Acien, F.G., Fernandez, J.M., and Molina, E. (2005). Modeling of growth and accumulation of carotenoids in *Haematococcus pluvialis* as a function of irradiance and nutrients supply. *Biochemical Engineering Journal*, 26: 107–114.

García, M.L., Del Río Sánchez, E., Casas López, J.L., Acién, F.G., Fernández, J.M., Fernandez Sevilla, J.M., Rivas, J., Guerrero, M.G., and Molina Grima, E. (2006). Comparative analysis of the outdoor culture of *Haematococcus pluvialis* in tubular and bubble column photobioreactors. *Journal of Biotechnology*, 29: 329–342.

Gladue, R.M., and Maxey, J.E. (1994). Microalgal feeds for aquaculture. *Journal of Applied Phycology,* 6: 131–141.

Gómez, P.I., Barriga, A., Silvia Cifuentes, A., and González, M.A. (2003). Effect of salinity on the quantity and quality of carotenoids accumulated by *Dunaliella salina* (strain CONC-007) and *Dunaliella bardawil* (strain ATCC 30861) Chlorophyta. *Biological Sciences,* 36: 185–192.

Gouveia, L., and Olieveira, A.C. (2009). Microalgae as a raw material for biofuel production. *Journal of Industrial Microbiology and Biotechnology,* 36: 269–274.

Gouveia, L., Marques, A.E., Da Silva, T.L., and Reis, A. (2009). *Neochloris oleoabundans* UTEX#1185: A suitable renewable lipid source for biofuel production. *Journal of Industrial Microbiology and Biotechnology,* 36: 821–826.

Gressel, J. (2008). Transgenics are imperative for biofuel crops. *Plant Science,* 174: 246–263.

Grima, E.M., Belarbia, E.H., Acieen Fernandeza, F.G., Robles Medina, A., and Chistib, Y. (2003). Recovery of microalgal biomass and metabolites: Process options and economics. *Biotechnology Advances,* 20: 491–515.

Grobbelaar, J.U. (2009). Factors governing algal growth in photobioreactors: The "open" versus "closed" debate. *Journal of Applied Phycology,* 21: 489–492.

Grobbelaar, J.U. (2010). Microalgal biomass production: challenges and realities. *Photosynthesis Research,* 106: 135–144.

Haag, A.L. (2007). Algae bloom again. *Nature,* 447: 520–521.

Harel, Y., Ohad, I., and Kaplan, A. (2004). Activation of photosynthesis and resistance to photoinhibition in cyanobacteria with biological desert crust. *Plant Physiology,* 136: 3070–3079.

Harun, R., Singh, M., Forde, G.M., and Danquah, M.K. (2010). Bioprocess engineering of microalgae to produce a variety of consumer products. *Renewable and Sustained Energy Reviews,* 14: 1037–1047.

Hejazi, M.A., Holwerda, E., and Wijffels, R.H. (2004). Milking microalga *Dunaliella salina* for β-carotene production in two-phase bioreactors. *Biotechnology and Bioengineering,* 85: 475–481.

Heredia-Arroyo, T., Wei, W., and Hu, B. (2010). Oil accumulation via heterotrophic/mixotrophic *Chlorella protothecoides.* *Applied Biochemistry and Biotechnology,* 162: 1978–1995. http://techtransfer.universityofcalifornia.edu/NCD/21141.html. A Hybrid Pond-Bioreactor for Mass Algal Culture. Tech ID: 21141/UC Case 2010-280-0.

Huerlimann, R., Nys, R., and Heimann, K. (2010). Growth, lipid content, productivity, and fatty acid composition of tropical microalgae for scale-up production. *Biotechnology and Bioengineering,* 107: 245–257.

Huesmann, M.H. (2000). Can advances in science and technology prevent global warming? A critical review of limitations and challenges. *Mitigation and Adaptation Strategies for Global Change,* 11: 539–577.

Hunt, R.W., Chinnaswamy, S., Bhatnager, A., and Das, K.C. (2010). Effect of biochemical stimulants on biomass productivity and metabolite content of microalgae, *Chlorella sorokiniana. Applied Biochemistry and Biotechnology,* doi: 10.1007/s2010-010-9012-2.

Huntley, M.E., and Redalje, D.G. (2007). CO_2 mitigation and renewable oil from photosynthetic microbes: A new appraisal. *Mitigation and Adaptation Strategies for Global Climate Change,* 12: 573–608.

Janssen, M., Tramper, J., Mur, L.R., and Wijffels, R.H. (2003). Enclosed outdoor photobioreactors: Light regime, photosynthetic efficiency, scale-up, and future prospects. *Biotechnology and Bioengineering,* 81: 193–210.

Jin, E., Feth, B., and Melis, A. (2003). A mutant of the green alga *Dunaliella salina* constitutively accumulates zeaxanthin under all growth conditions. *Biotechnology Bioengineering,* 81: 116–124.

Johnson, M.B. (2009). Microalgal Biodiesel Production through a Novel Attached Culture System and Conversion Parameters. M.Sc. thesis in Biological Systems Engineering, Blacksburg, VA, p. 83.

Kaeriyama, H., Katsuki, E., Otsubo, M., Yamada, M., Ichimi, K., Tada, K., and Harrison, P.J. (2011). Effects of temperature and irradiance on growth of strains belonging to seven *Skeletonema* species isolated from Dokai Bay, southern Japan. *European Journal of Phycology,* 46: 113–124.

Kanellos, M. (2009). Algae Biodiesel: It's $33 a Gallon. Greentech Media. [Online] Greentech Media, 03 02 2009. [Cited: 10 12 2009.] http://www.greentechmedia.com/articles/algae-biodiesel-its-33-a-gallon-5.

Kebede-Westhead, E., Pizarro, C., and Mulbry, W.W. (2006). Treatment of swine manure effluent using freshwater algae; Production, nutrient recovery, and elemental composition of algal biomass at four effluent loading rates. *Journal of Applied Phycology,* 18: 41–46.

Keerthi, S., Uma Devi, K., Subba Rao, D.V., Sarma, N.S. (in press). Exogenous vitamin dependency in two carotenogenic *Dunaliella* strains isolated from coastal Bay of Bengal.

Khan, S.A., Hussain, M.Z., Prasad, S., and Banerjee, U.C. (2009). Prospects of biodiesel production from microalgae in India. New Delhi. India 2009. *Renewable and Sustainable Energy Reviews,* 13: 2361–2372.

Khozin-Goldberg, I., and Cohen, Z. (2006). The effect of phosphate starvation on the lipid and fatty acid composition of the freshwater eustigmatophyte *Monodus subterraneus. Phytochemistry,* 67: 696–701.

Kleinegris, D.M.M., Janssen, M., Brandenburg, W.A., and Wiiffels, R.H. (2011). Two-phase systems: Potential for in situ extraction of microalgal products. *Biotechnology Advances,* 29: 502–507.

Kong, Q.X., Li, L., Martinez, B., Chen, P., and Ruan, R. (2010). Culture of microalgae *Chlamydomonas reinhardtii* in wastewater for biomass feedstock production. *Applied Biochemistry and Biotechnology,* 160: 9–18.

Lebeau, T., and Robert, R.M. (2006). Biotechnology of immobilized microalgae. A culture technique for future? In *Algal Cultures, Analogues of Blooms and Applications* (Ed. Subba Rao, D.V.), Science Publishers, Enfield, NH, pp. 801–839.

Lee, Y.K., Ding, S.Y., Hoe, C.H., and Low, C.S. (1996). Mixotrophic growth of *Chlorella sorokiniana* in outdoor enclosed photobioreactor. *Journal of Applied Phycology,* 8: 163–169.

León, R., and Fernández, E. (2007). Nuclear transformation of eukaryotic microalgae. In *Transgenic Microalgae as Green Cell Factories* (Eds. Leon, R., Galvan, A., and Fernandez, E.). Landes Bioscience and Springer Science+ Business Media, New York, pp. 1–11.

Li, S., and Tsai, H. (2008). Transgenic microalgae as a non-antibiotic bactericide producer to defend against bacterial pathogen infection in the fish digestive tract. *Fish & Shellfish Immunology,* 26: 316–325.

Li, Y., Horsman, M., Wang, B., Wu, N., and Lan, C.Q. (2008). Effects of nitrogen sources on cell growth and lipid accumulation of green alga *Neochloris oleoabundans. Applied Microbiology and Biotechnology,* 81: 629–636.

Li, H., Xu, H., and Wu, Q. (2007). Large-scale biodiesel production from microalga *Chlorella protothecoides* through heterotrophic cultivation in bioreactors. *Biotechnology and Bioengineering,* 98: 764–771.

Liu, Z.Y., Wang, G.C., and Zhou, B.C. (2008). Effect of iron on growth and lipid accumulation in *Chlorella vulgaris. Bioresource Technology,* 99: 4717–4722.

Major, K.M., and Henley, W. (2008). Influence of salinity and temperature on growth and photosynthesis in the extremophilic chlorophyte, *Nannochloris* sp. *Journal of Phycology,* 37: 32.

Malcata, F.X. (2010). Microalgae and biofuels: A promising partnership? *Trends in Biotechnology,* 29: 542–559.

Mandal, S., and Mallick, N. (2009). Microalga *Scenedesmus obliquus* as a potential source for biodiesel production. *Applied Microbiology and Biotechnology,* 84: 281–291.

Mata, T.M., Martins, A.A., and Caetanao, N.S. (2010). Microalgae for biodiesel production and other applications: A review. *Renewable and Sustainable Energy Reviews,* 14: 217–232.

Matsumoto, M., Sugiyama, H., Maeda, Y., Sato, R., Tanaka, T., and Matsunaga, T. (2009). Marine diatom, *Navicula* sp. strain JPCC DA0580 and marine green alga, *Chlorella* sp. strain NKG400014 as potential sources for biodiesel production. *Applied Biochemistry and Biotechnology,* 16: 483–490.

McGinnis, K.M., Dempster, T.A., and Sommerfeld, M.R. (1997). Characterization of the growth and lipid content of the diatom *Chaetoceros muelleri. Journal of Applied Phycology,* 9: 19–24.

Miquel, P. (1893). De la culture artificielle des diatoms. Introduction. *Le Diatomiste,* 1: 73–75.

Misumi, O., Yoshida, Y., Nishida, K., Fujiwara, T., and Sekajin, T. (2008). Genome analysis and its significance in four unicellular algae, *Cyanidioshyzon merolae, Ostreococcus tauri, Chlamydomonas reinhardtii,* and *Thalassiosira pseudonana. Journal of Plant Research,* 121: 3–17.

Mitra, M., and Melis, A. (2008). Optical properties of microalgae for enhanced biofuels production. *Optics Express,* 16: 21807–21820.

Morgan-Kiss, R.M., Ivanov, A.G., Modia, S., Czymmek, K., and Huner, N.P. (2008). Identity and physiology of a new psychrophilic eukaryotic green alga, *Chlorella* sp. strain BI, isolated from a transitory pond near Bratina Island, Antarctica. *Extremophiles,* 12: 701–711.

Muradyan, E.A., Klyachko-Gurvich, G.L., Tsoglin, L.N., Sergeyenko, T.V., and Pronina, N.A. (2004). Changes in lipid metabolism during adaptation of the *Dunaliella salina* photosynthetic apparatus to high CO_2 concentration. *Russian Journal of Plant Physiology,* 51: 53–62.

Natural Resources Defense Council (2009). *The Promise of Algae Biofuels.* Catie Ryan, Terrapin Bright Green, LLC, New York, p. 81.

Norsker, N., Barbosa, M.J., Vermue, M.H., and Wijffels, R.H. (2011). Microalgal production – A close look at the economics. *Biotechnology Advances,* 29: 24–27.

Olaizola, M. (2000). Commercial production of astaxanthin from *Haematococcus pluvialis* using 25,000-liter outdoor photobioreactors. *Journal of Applied Phycology,* 12: 499–506.

Orpez, R., Martinez, M.E., Hodaifa, G., Yousfi, F. El., Sanchez, S., and Jbari, N. (2009). Growth of the microalga *Botryococcus braunii* in secondarily treated sewage. *Desalination,* 246: 625–630.

Park, J.B.K., Craggs, R.J., and Shilton, A.N. (2011). Wastewater treatment high rate algal ponds for biofuel production. *Bioresource Technology,* 102: 71–81.

Perez-Garcia, O., De-Basham, L., Hernandez, J., and Bashan, Y. (2010). Efficiency of growth and nutrient uptake from wastewater by heterotrophic, autotrophic, and mixotrophic cultivation of *Chlorella vulgaris* with *Azospirillum brasiliense. Phycology,* 46: 800–812.

Pienkos, P.T., and Darzins, A. (2009). The promise and challenges of microalgal-derived biofuels. *Biofuels, Bioproducts and Biorefining,* 3: 431–440.

Pienkos, P.T., Laurens, L., and Aden, A. (2011). Making biofuel from microalgae. *American Scientist,* 99: 474–481.

Pitman, J.K., Dean, A.P., and Osundeko, O. (2011). The potential of sustainable algal biofuel production using wastewater resources. *Bioresource Technology,* 102: 17–25.

Pocock, T., Vetteril, A., and Falk, S. (2011). Evidence of phenotypic plasticity in the Antarctic extremophile *Chlamydomonas raudenis* Ettl.UWO 241. *Journal of Experimental Botany,* 62: 1169–1177.

Pruvost, J., Van Vooren, G., and Legrand, C. (2009). Investigation of biomass and lipids production with *Neochloris oleoabundans* in photobioreactor. *Bioresource Technology,* 100: 5988–5995.

Pulz, O. (2007). Performance Summary Report: Evaluation of Green Fuel's 3D Matrix Algae Growth Engineering Scale Unit, APS Red Hawk Power Plant, AZ. IGV Instüt für Getreideverabeitung GmbH, pp. 1–14.

Pulz, O., and Gross, W. (2004). Valuable products from biotechnology of microalgae. *Applied Microbiology and Biotechnology,* 65: 635–648.

Radakovits, R., Jinkerrson, R.E., Darzins, A., and Posewitz, M.C. (2010). Genetic engineering of algae for enhanced biofuel production. *Eukaryotic Cell,* 9: 486–501.

Raven, J.A. (2009). Carbon dioxide fixation by *Dunaliella* spp. and the possible use of this genus in carbon dioxide mitigation and wastewater reduction. In *The Alga Dunaliella Biodiversity, Physiology, Genomics and Biotechnology* (Eds. Ben-Amotz, A., Polle, J.E.W., and Subba Rao, D.V.), Science Publishers, Enfield, NH, pp. 359–384.

Riegman, R., Boer, M., and Senerpont Domis, L. (1996). Growth of harmful marine algae in multispecies cultures. *Journal of Plankton Research,* 18: 1851–1866.

Rivasseau, C., Farhi, E., Gromova, M., Oliver, J., and Bligny, R. (2010). Resistance to irradiation of micro-algae growing in the storage pools of a nuclear reactor investigated by NMR and neutron spectroscopies. *Spectroscopy: An International Journal,* 24: 381–385.

Rodolfi, L., Zitteli, G.C., Bassi, N., Padovani, G., Biondi, N., Bonini, G., and Tredici, M.R. (2008). Microalgae for oil: Strain selection, induction of lipid synthesis and outdoor mass cultivation in a low-cost photobioreactor. *Biotechnology and Bioengineering,* 102: 100–112.

Ronquillo, J.D., Matias, J.R., Saisho, T., and Yamasaki, S. (1997). Culture of *Tetraselmis tetrathele* and its utilization in the hatchery production of different penaeid shrimps in Asia. *Hydrobiologia,* 358: 237–244. doi: 10.1023/A:1003128701968.

Rosenberg, J.N., Oyler, G.A., Wilkinson, L., and Betenbaugh, M.J. (2008). A green light for engineered algae: redirecting metabolism to fuel a biotechnology revolution. *Biotechnology,* 19: 430–436.

Russell, G., and Fielding, A.H. (1974). The competition properties of marine algae. *Journal of Ecology,* 62: 689–698.

Scott, S.A., Davey, M.P., Dennis, J.S., Horst, I., Howe, C.J., and Smith, A.G. (2010). Biodiesel from algae: challenges and prospects. *Current Opinion Biotechnology,* 21: 277–286.

Serrano, L. (2010). Five hard truths for synthetic biology. *Nature,* 463: 288–290.

Service, R.F. (2008). Eyeing oil, synthetic biologists mine microbes for black gold. *Science,* 322: 522–523.

Spijkerman, E. (2010). High photosynthetic rates under a co-limitation for inorganic phosphorus and carbon dioxide. *Journal Phycology,* 46: 658–664.

Spijkeman, E., and Wacker, A. (2011). Interactions between P-limitation and different C conditions on the fatty acid composition of an extremophile microalga. *Extremophiles,* 15: 597–609.

St. John, J. (2009). Algae Company Number 56: Plankton Power. Greentech Media. [Online] Greentech Media Inc, 04 08 2009. [Cited: October 12 2009.] http://www.greentechmedia.com/articles/read/plankton-power-another-algae.

Stephens, E., Ross, I.L., Mussgnug, J.H., Wagner, L.D., Borowitzka, M.A., Posten, C., Kruser, O., and Hankamer, B. (2010). Future prospects of microalgal biofuel production systems. *Trends in Plant Science,* 15: 554–564.

Sturm, B.S.M., and Lamer, S.L. (2011). An energy evaluation of coupling nutrient removal from wastewater with algal biomass production. *Applied Energy,* 88: 3499–3506.

Subba Rao, D.V. (2009). Cultivation, growth media, division rates and applications of *Dunaliella* species. In *The Alga Dunaliella Biodiversity, Physiology, Genomics and Biotechnology* (Eds. Ben-Amotz, A., Polle, J.E.W., and Subba Rao, D.V.). Science Publishers, Enfield, NH, pp. 45–90.

Subba Rao, D.V., Pan Y., and Al-Yamani, F. (2005). Growth and photosynthetic rates *of Chalmydomonas plethora* and *Nitzschia frustula* cultures isolated from Kuwait Bay, Arabian Gulf and their potential as live algal food for tropical mariculture. *Marine Ecology,* 26: 63–71.

Takagi, M., Karseno, S., and Yoshida, Y. (2006). Effect of salt concentration on intracellular accumulation of lipids and triglyceride in marine microalgae *Dunaliella* cells. *Journal of Bioscience and Bioengineering,* 101: 223–226.

Theegala, C.S., Robertson, C., and Suleiman, A.A. (2001). Phytoremediation potential and toxicity of barium to three freshwater microalgae: *Scenedesmus subspicatus, Selenastrum capricornutum,* and *Nannochloropsis* sp. *Practice Periodical of Hazardous, Toxic, and Radioactive Waste Management,* 5: 194–202.

Tsoglin, L.N., and Gabel, B.V. (2000). Potential productivity of microalgae in industrial photobioreactor. *Russian Journal of Plant Physiology,* 47: 668–673.

Ugwu, C.U., Aoyagi, H., and Uchiyama, H. (2008). Photobioreactors for mass cultivation of algae. *Bioresource Technology,* 99: 4021–4028.

Wahal, S., and Viamajala, S. (2010). Maximizing algal growth in batch reactors using sequential change in light intensity. *Applied Biochemistry and Biotechnology,* 161: 511–522.

Weber, A.P.M., Oesterhelt, C., Gross, W., Brautigam, A., Imboden, L.A., Krassovskayal, I., Linka, N., Truchina, J., Zimmermann, M., Jamai, A., Riekhof, W.R., Yu, B., Garavito, R.M., and Benning, C. (2004). EST—Analysis of the thermo-acidophilic red microalga *Galderia sulphuraria* reveals potential for lipid A biosynthesis and unveils the pathway of carbon export from rhodoplasts. *Plant molecular Biology,* 55: 17–32.

Wang, L., Li, Y., Chen, P., Min, M., Chen, Y., Zhu, J., and Ruan, R.R. (2010). Anaerobic digested dairy manure as a nutrient supplement for cultivation of oil-rich green microalgae *Chlorella* sp. *Bioresource Technology,* 101: 2623–2688.

Weyer, K.M., Bush, D.R., Darzins, A., and Willson, D.B. (2010). Theoretical maximum algal oil production. *Bioenergy Research,* 3: 204–213.

Wijffels, R.H. (2007). Potential of sponges and microalgae for marine biotechnology. *Trends in Biotechnology,* 26(1): 26–31.

Williams, P.J. le B., and Laurens L.M.L. (2010). Microalgae as biodiesel and biomass feedstocks. *Energy and Environmental Science,* 3: 554–590.

Woertz, I., Feffer, A., Lundquist, T., and Nelson, Y. (2009). Algae grown on dairy and municipal wastewater for simultaneous nutrient removal and lipid production for biofuel feedstock, *Journal of Environmental Engineering,* 135: 1115–1122.

Wu, X., and Merchuk, J.C. (2004). Simulation of algae growth in a bench scale internal loop airlift reactor. *Chemical Engineering Science,* 59: 2899–2912.

Xu, H., Miao, X.L., and Wu, Q.Y. (2006). High quality biodiesel production from a microalga *Chlorella protothecoides* by heterotrophic growth in fermenters. *Journal of Biotechnology,* 126: 499–507.

Index